GROVE'S FIELD GUIDE TO PLASTICS

Unpacking the Plastics Pandemic

Bob Grove

Wordsinger Publishing

The information in this book was correct at the time of publication. The Author does not assume any liability for loss or damage caused by errors or omissions.

ISBN: 978-1-961014-06-0
LCCN: 2024903534

Published in the United States of America

Wordsinger Publishing
Hilo, Hawaii
www.Wordsinger.com

CONTENTS

INTRODUCTION

Where's the plastic? It's everywhere. It's on the ground and in our oceans. It's in the air, in our food and in our water. It's in our guts, lungs, and blood streams. It's in placentas and breast milk. We breathe it. We drink it. We eat it and we wear it.

This one really snuck up on us. One moment we're revolutionizing the future and the next we're swimming in a forever land of nondegradable microplastic particles, fibers and the cornucopia of toxic forever chemicals used to make them. Plastics are complex petroleum-based carbon chains with the same climate impacts as burning coal and gasoline, and a centuries-long life span that will leave them clogging our waterways and landfills centuries after we're gone. They never break down into their component chemicals; they just get brittle and break apart into smaller and smaller pieces of – well, plastic.

The health problems of having hundreds of millions of tiny plastic particles taking up space inside our bodies is just now being studied. Early research on the chemicals leaching from these microplastics suggest links to ADHD, obesity, Alzheimer's and a pantheon of other neurological and hormonal issues.

We can't hide from this modern world of plastic pollution, but we can unpack the problem and choose to avoid the greatest risks. It's our responsibility to make personal choices that assure the health, safety and welfare of ourselves, our communities, our children and future generations. And that's where this Field Guide will help.

> *TIP: This introduction is going to talk about science, history and politics. If you'd prefer to get right to the problems and solutions, feel free to skip to the next chapter.*

The science behind plastics is pretty simple if we ignore the politics. We've been using plastic-like materials for ages.[1] Mesoamericans were using sap from the rubber tree (*Hevea brasiliensis*) to create balls and figurines in 1600 BCE. Shellac, made from the secretion of the "lac" insect (*Kerria lacca*), was around in 1000 BCE. Cow and ox horn were softened to make windows and lamp shades throughout the middle ages.

In the 1800s, we started taking these naturally flexible things and chemically modifying them to produce things NOT found in nature. Polystyrene was created in 1839 by Eduard Simon, a German apothecary. Vulcanized rubber was patented in 1844 by Thomas Hancock in Britain and Charles Goodyear in the United States. Parkesine, the first celluloid, was patented – and humbly named after himself – in 1856 by Alexander Parkes. PVC was discovered in 1869 by German chemist Eugen Baumann. Celluloid film was patented in 1889 by Eastman Kodak. Galalith was developed in 1890 by Wilhelm Krische and Adolph Spitteler. Shellac phonograph records became industry standard in 1894. And polyethylene was synthesized in 1898 by German chemist Hans von Pechmann.

In the 1900s, we ditched nature and started developing completely synthetic products. Bakelite, the world's first completely synthetic plastic, was developed in 1907 by Leo Baekeland. PVC was "synthesized" in 1926 by Waldo Semon at the B.F. Goodrich Company. Neoprene was produced in 1930 by DuPont. Polystyrene in 1930 by BASF. Vinyl "Victrolac" records were introduced in 1931 by RCA Victor. Polyethylene was synthesized in 1933 by the Imperial Chemical Industries. Nylon was invented by Wallace Carothers and patented in 1935 by DuPont. Teflon was discovered in 1938 by Roy Plunkett at DuPont. PET (Polyethylene terephthalate) was discovered in 1941 at the Calico Printers' Association. Polyester was manufactured in 1950 by DuPont. Polypropylene produced in 1951 by J. Paul Hogan and Robert L. Banks from at Phillips. Expanded polystyrene was invented in 1954 by Dow Chemical. High-density polyethylene bottles were introduced in 1960 and soon replaced glass bottles. Kevlar was developed in 1965 by Stephanie Kwolek at DuPont. Polyester film stock replaced cellulose acetate in 1980 and the first plastic banknotes were issued in 1988 by Australia.

> *TRIVIA: The word "plastic" has been around for centuries. It's derived from the Greek πλαστικός (plastikos) and is a synonym for the word malleable, meaning "capable of being shaped or formed." The modern term "plastics" was coined by Leo Baekeland in 1907 when he created Bakelite, the world's first fully synthetic polymer. Leo combined a coal tar extract named phenol (C6H5OH) with formaldehyde (CH2O) to create his new plasticized resin (C6H6O-CH2OH) which he named after himself.*

Plastics have a long history with lots of apothecaries and chemists in it, which isn't surprising since plastics and their additives are all about the chemicals. And let's face it. If we want to understand the stuff piling up in our backyards, we're going to have to talk some chemistry.

But don't worry. Chemistry's really not that hard. We already know most of the words. We've heard of elements, atoms, molecules and compounds. We just need to understand how they relate to each other. An atom is the smallest possible piece of a chemical element.[2] Combine two or more atoms together and you have molecules.[3] Combine different molecules together and you have compounds.[4] Some compounds are monomers[5], which just means that they can connect to each other. Connect together long repeating chains of monomers and you have polymers.[6] Done.

> *TRIVIA: The word "polymer", is derived from the Greek πολύς (polus) 'many', and μέρος (meros) 'part'.*

All plastic-like substances, both natural and synthetic, are made of polymers. Our DNA is a polymer – long repeating chains of nucleotides. The proteins in our food are polymers – long, repeating chains of amino acids. Latex, the sap from the rubber tree, is a natural polymer – long, repeating chains of isoprene ($CH_2=C(CH_3)-CH=CH_2$). Styrofoam® is a synthetic polymer – long, repeating chains of styrene ($C_6H_5CH=CH_2$). Kevlar is a synthetic polymer – long, repeating chains of ... well, $-CO-C_6H_4-CO-NH-C_6H_4-NH-$. Polymers are a major field of study, spanning polymer science, biophysics, materials science and engineering. Chemical and plastics companies are all about polymer science, both polymer chemistry and polymer physics.

There are currently 118 known elements and the study of chemistry is really just about finding new ways to combine these 118 elements into new molecules and compounds. Chemical formulas are a chemist's shorthand way to describe all the different ways these 118 elements can be combined. In formulas, the elements are abbreviated as one or two letters with the number of atoms of each element written as a subscript after the letter. For instance, common elements we see a lot of are:

C = Carbon
H = Hydrogen
O = Oxygen
N = Nitrogen

The chemical formula for a molecule of water is H_2O. That's two hydrogen atoms with one oxygen atom. The chemical formula for a molecule of Tryptophan (one of the essential amino acids in our food) is $C_{11}H_{12}N_2O_2$. Yep, that's eleven atoms of carbon, 12 atoms of hydrogen, two nitrogen and two oxygen. You're now an official amateur chemist.

TIP: Fossil fuels are all hydrocarbons. That's hydro (hydrogen) + carbon. So expect to see a lot of Hs and Cs in plastic formulas.

The other thing we want to understand is the chemical bonds that hold molecules together. This helps us understand why plastics last for centuries. Atoms are composed of just three things – protons, neutrons and electrons. Protons and neutrons are particles that sit in the center of an atom – an area called the nucleus. Electrons are particles that orbit around the nucleus. We've all seen pictures of our solar system, with its 8-9 planets orbiting the sun. An atom is kind of like a super-tiny solar system with its electrons orbiting its nucleus. That's it. Now you're an amateur physicist.

TRIVIA: Our formerly ninth planet, Pluto, was relegated to a mini-planet in 2006 by the International Astronomical Union when they redefined the meaning of planet to be a celestial body that 1) orbits the sun, 2) is round, and 3) "clears the neighborhood" around its orbit. Pluto failed test three because it was part of the Kupier Belt of objects and because its orbit overlapped Neptune's. However, when NASA's New Horizons spacecraft reached Pluto in 2015 and found it and its moons to be larger and more complex than imagined, the planetary status debate was reignited.

You can identify an element by how many protons its atom has in its nucleus. This is referred to as its atomic weight. A hydrogen atom has one proton in its nucleus, so its atomic weight is one. Carbon has six protons, nitrogen has seven and oxygen has eight – all the way up to the element Oganesson which has 118 protons in its nucleus. This is what's on that big *periodic table* hanging in most science classrooms.[7]

> *TRIVIA: Of course, the element Oganesson was discovered by the Russian physicist Yuri Oganessian and named after himself.*

There are different types of chemical bonds, but the one that applies to plastics is the covalent chemical bond.[8] This is created by the electrons orbiting around an atom's nucleus. Some atoms don't have enough electrons to be stable, so they "borrow" electrons from atoms near them. When two atoms keep "borrowing" the same electron from each other, the electron finds itself in orbit around both atom's nuclei, bonding them together. Imagine that tiny atomic solar system with an electron orbiting around its nucleus. Now imagine that electron in a figure eight orbit around two atoms' nuclei. You're imagining a covalent chemical bond. The atoms can't move away from each other without giving up their shared electron. They're chemically bonded. Now imagine atoms sharing two electrons. This is a double bond. And imagine atoms sharing THREE electrons. This is the super strong triple bond!

Covalent bonds are written in structural formulas[9] using lines. They tell us the structure of the molecule, not just its molecular formula. One line (-) represents a weak single bond. Two lines (=) represent a double bond. And three lines (≡) represent a strong triple bond. For instance, styrene's structural formula is $C_6H_5CH{=}CH_2$. This tells us that there's a double bond (two shared electrons) between the C_6H_5CH molecules and the CH_2 molecule.

> *TRIVIA: You may be asking yourself why C6H5CH isn't written C7H6. That's because C6H5 and CH are separate molecules with a single bond between their carbon atoms. Chemists like to omit the single bonds to keep formulas short. Both C6H5-*

CH=CH2 and C6H5CH=CH2 are legitimate ways of saying styrene.

The path from fossil fuel to plastic is complex but predictable. A fossil fuel is extracted from the ground, refined into *fractions* that are cracked into monomers which are chemically bonded into polymers.[10] The polymers are then compounded with any of 10,000+ different chemical additives[11] to create different types of plastic resin which are processed into *nurdles*[12] and shipped to factories around the world where they are melted and turned into the cornucopia of plastics we tragically love. *Whew!*

TRIVIA: Nurdles are tiny pellets of plastic resin, smaller than 0.20 inches (5mm).

Pathway from Fossil Fuel to Plastic

Extract – Extract it from the earth.
Refine – Refine it into its component fractions.
Crack – Break the fractions into monomers.
Polymerize – Bond the monomers into polymer chains.
Compound – Add chemicals to create resins.
Process – Cut the resins into nurdles.
Ship – Transport the nurdles to factories.
Manufacture – Melt and shape nurdles into products.

A *fraction* is a unique part of a raw fossil fuel. Different fractions have different boiling points, so when crude oil, for example, is vaporized and pumped into a fractional distillation tower[13] to cool, its fractions separate. The tower is hottest at the bottom and coldest at the top so, as they rise, fractions condense at different levels of the tower and run off into storage. The lightest fractions with the lowest boiling points condense at the top of the tower where it's coolest, and the heaviest fractions with the highest boiling points condense at the bottom. Natural gas dissolved in the crude oil exits the top of the tower and moves on to a gas refinery.[14]

Distillation Tower Fractions
Boiling Point Range – Product (use)

< 85°F – Refinery Gases (butane, bottled gas, PLASTICS)
85-135°F – Gasoline
135-350°F – Naphtha (chemicals, PLASTICS)
350-450°F – Kerosine (jet fuel, heating)
450-650°F – Diesel
650-1050°F – Heavy Fuel Oil (lubricants, ships)
> 1050°F – Residue Oil (asphalt, roofing)

The naphtha fraction from our crude oil refinery is cracked into ethylene (C_2H_4 or $H_2C=CH_2$) that is polymerized into polyethylene ($\sim CH_2CH_2CH_2CH_2CH_2CH_2\sim$), compounded with chemical additives to make resin types #1, #2 & #4, processed into nurdles and shipped to factories where the nurdles are melted and manufactured into things like flexible food containers, water bottles, milk jugs, plastic bags and carpet.[15]

Natural gas – either from the top of our crude oil fractional distillation tower or fracked from a gas well – is run through a series of distillation columns[16] where pressure and cryogenic refrigeration is used to condense the gas into its fractions. The propane fraction (C_3H_8 or $CH_3CH_2CH_3$) is cracked into the monomer propylene ($CH_2=CHCH_3$), polymerized into polypropylene ($\sim C_3H_6C_3H_6C_3H_6C_3H_6C_3H_6C_3H_6\sim$), compounded with chemical additives into resin type 5, processed into nurdles and shipped to factories where the nurdles are melted and manufactured into things like auto parts, rigid food containers and dishware.

Even coal can be made into plastic. Coal is mined then liquified into methanol,[17] which can be cracked into both ethylene and propylene monomers, which then follow the paths-to-plastic described above.

All paths-to-plastic consume lots of energy & water, release huge amounts of greenhouse gas, and produce more than their fair share of toxic wastes. You won't be surprised to learn that the coal-to-plastic path has the biggest environmental impact,[18] with three times the carbon emissions of oil-to-plastic and 4.2 times the water consumption.[19]

In 2019, synthetic plastic production emitted 850 million metric tons of greenhouse gases. That's a lot. It's the equivalent of 185 new 500-megawatt coal power plants. By 2030, emissions are predicted to be 1.32 billion metric tons per year – more than 295 new 500-megawatt coal-fired power plants.[20]

Another big problem with synthetic plastics is that they don't decompose. They're made by breaking fossil fuels down and reassembling them into compounds that have never existed on earth, and these new compounds are alien to the earth's decomposition cycle.[21] There's just nothing in nature that can figure out how to take them apart. So the plastic polymers and their chemical additives last for centuries. But we'll go into that more later.

Once Again, With Politics

That was a lot of science stuff. We obviously have a very clear understanding of the processes and problems. So why don't we fix things? Well, let's bring politics back into this for a moment.

A lot of very fundamental things changed after World War II. Things like how we farm, the way we eat, and the creation of retail consumerism[22] and single-use products[23]. President Franklin D Roosevelt established the War Production Board In 1942.[24] By funneling military money into civilian industry, this agency was able to expand the production of military products 4000%, thus creating the US Military-Industrial Complex.[25]

The use of synthetic plastics surged 300% during World War II.[26] Military research created innovative food packaging for the soldiers, lightweight windshields for war planes, parachutes, etc. And the expansion of plastics continued after the war. Military food products were re-branded and sold to the American home as processed foods[27] along with the new plastic appliances to prepare them. Military food packaging became Baggies® and Tupperware®. Parachutes became nylon stockings. Airplane windshields became Plexiglas® furniture, shower doors, etc.[28]

TRIVIA: Processed foods from World War II have become so common that NOAA found a can of SPAM® in the Marianas Trench 16,473' under water.

TRIVIA: The ammunition industry converted to fertilizer production, leading to the petrolification of agriculture. They promoted a monocropping farm system that relies on chemical herbicides, pesticides and fertilizers to raise the same crop year after year. Notice how similar nitrogen fertilizers (CO(NH2)2) are to nitroglycerin (C3H5N3O9).

Since the war, the fossil fuel industry has figuratively exploded to keep up with the growth of nitrogen based (fossil fuel) fertilizers, power production, transportation and plastics.[29] And, with the rise of cost-effective renewable energy and electric cars, the industry is depending more and more on plastics. The International Energy Agency forecasts that plastics will consume 45% of all oil production by 2040. British Petroleum forecasts 95%.[30] Fossil fuel companies are planning to invest $400 billion into new petrochemical plants. Since 2010, they've already invested over $200 billion in 333 new plastic and other chemical projects in the US.[31] Shell Chemical has built a $6 billion ethane cracker thirty miles northwest of Pittsburgh.[32] Exxon

Mobil Corp has launched $20 billion in new refining and chemical manufacturing projects along the US Gulf Coast.[33]

In 2023, petroleum's global revenue was nearly three trillion dollars a year.[34] Is it any surprise that the industry might feel some pressure to continue making that kind of money? Is it any surprise that politicians might be inclined to help?[35]

> *TRIVIA: It took the US nearly two centuries to accumulate a $33 trillion national debt. The petroleum industry makes that much in a decade.*

With all the money that's at stake, it's understandable that there's pressure to keep this gusher going. And the fossil fuel industry has a track record of putting profits before people. They discovered global warming back in 1959 and kept it quiet.[36] When it finally hit the news in the 1980s, they drew from the Tobacco Industry Playbook[37] – which protected tobacco revenues by denying the link between smoking and cancer – to successfully sow doubt about climate change and to reposition global warming as a theory, not a science-backed fact. Now, they're using the same playbook to sow doubt about the risks of plastics.[38]

Turns out that they've known that plastic additives were building up in our blood and having toxic effects on our organs, for decades. Evidence from *State of Minnesota v. 3M Co.*[39] (Court File No. 27-CV-10-28862) show that 3M and DuPont were aware of the risks of plastic additives like PFAS as early as the 1950s.[40] A 3M study in the 1950s showed that PFAS build up in blood. Animal studies in the 1960s at both 3M and DuPont reveal that PFAS posed health risks. 3M was discussing the buildup of PFAS in Americans' blood in the 1970s. And studies at both 3M and DuPont in the 1980s linked PFAS to elevated cancer rates among their own employees.[41] A 1975 document states, "DuPont is concerned with the possible toxic effects of FC-143 in the Teflon K used in the Harshaw 'Dustless Process' because of its possible use in, or in contact with, food products. [H.M.] emphasized that duPont [*sic*] is not advocating the use of Teflon K in food applications and actually recommends against such use. They do not plan to file for an FDA clearance for Teflon K. However, they do recognize that they cannot control all applications and they would like to have defensive information available to support their position in the event that a problem does arise."[42]

None of these internal studies were published in scientific journals or shared with regulators or employees. A independent 2023 study titled "The Devil They Knew: Chemical Documents Analysis of Industry Influence on PFAS Science" states, "Our review of industry documents shows that companies knew PFAS was 'highly toxic when inhaled and moderately toxic when ingested' by 1970, forty years before the public health community. Further, the industry used several strategies that have been shown common to tobacco, pharmaceutical and other industries to influence science and regulation – most notably, suppressing unfavorable research and distorting public discourse."[43]

A 2023 study of lobbying documents found that US chemical companies spent over $110 million from 2019-2022 to defeat PFAS legislation and slow administrative regulation of PFAS.[44]

The European Union (EU), a voluntary union of 27 independent nations, leads the world in testing and banning toxic chemical additives. The United States (US) takes a more corporate-friendly approach. The US Food and Drug Administration (FDA) often finds no evidence of human toxicity for substances flagged or banned by the EU. And the US Environmental Protection Agency (EPA) tends to encourage the voluntary phase-out of substances while they continue to perform risk assessments.

On the microplastics front, the industry is using the playbook to deflect concern about a 2024 Columbia University study published in the *Proceedings of the National Academy of Sciences*, which found an average of 250,000 nanoplastics in a standard bottle of water.[45] The International Bottled Water Association responded by questioning the testing methods, then urged people to stay calm and keep drinking bottled water while scientists do more research.[46] This is standard playbook language.

The Chemours Company – which was created in 2015 to shield DuPont from legal liability over damages caused by PFAS, titanium dioxide, and other chemicals[47] – reported that they had reduced PFAS air emissions at a North Carolina factory by 99.99%. However, when The Guardian News ran their own tests, they found PFAS levels 30 times higher than reported.[48] Chemours responded by questioning the testing methods and called the Guardian's findings "false" and "irresponsible". More playbook language.

Evidence is now emerging that the plastics industry has known about the plastic waste crisis for decades and promoted plastic recycling programs in a classic playbook tactic of deflect, distract and delay. The plastics industry established the Council for Solid Waste Solutions in 1988 – a group highly reminiscent of the Information Council for the Environment[49] that used the playbook to deny climate change. The job of this and other plastics trade organizations was to defend the plastics industry from legislation by selling recycling as the solution to plastic waste. As a DuPont director put it in 1988, "No doubt about it, legislation is the single most important reason why we are looking at recycling."[50]

In 2024, a report titled "The Fraud of Plastic Recycling: How Big Oil and the plastics industry deceived the public for decades and caused the plastic waste crisis"[51] was picked up by PBS,[52] NPR,[53] The Guardian[54] and The Hill.[55] This report cites a series of internal plastic industry letters, staff meeting notes and other documents that support the report's title argument. Among them, was a 1973 report prepared for The Society of the Plastics Industry that states, "There are no effective market mechanisms for trade in contaminated, mixed plastics. Sorting by the homeowner over any sufficiently practical period of time has been shown to be infeasible." A 1986 fact sheet written by the Vinyl Institute states, "Recycling cannot be considered a permanent solid waste solution, as it merely prolongs the time until an item is disposed of." A 1989 paper titled "Achieving Expansion Through Plastics Recycling" states, "Recycling cannot go on indefinitely, and does not solve the solid waste problem." A 1996 summary report for the Association of Plastics Manufacturers in Europe states, "Eco-impact studies in the Netherlands and Germany have demonstrated that there is a limit (18 per cent maximum in the Netherlands) to the amount of household plastics which can be mechanically recycled with environmental gain."

In classic playbook fashion, the plastics industry is now pushing a *circular economy*. Isn't that just a new name for recycling? Not only do their own internal studies say this doesn't work, but growing evidence shows that remolding used plastics create dioxins and other banned chemicals.[56] It's hard to be "circular" when parts of the circle are being banned.

Look, you really can't blame corporations for being what they are. They were created for the sole purpose of maximizing profits. Every measurement of corporate success, whether on the factory floor, in the board room or at the stock market, is a measurement of higher production and profit. So you can't blame corporations for wanting their products to be necessities of life. You certainly can't blame them for wanting to keep their coal and gas power plants running after all the money they invested in building them. And you can't blame them for wanting to continue drilling, fracking and mining after all the money they've invested in extraction and refineries. And you really can't blame corporations for redirecting their fossil fuels into the production of plastics now that renewable energy and electric vehicles are cutting into traditional profits.

We can all imagine a fossil fuel board meeting that goes,

Thanks, Tom. We're currently making what, $263 million a day off power plants? How long can we keep these babies running?

Our scientists are making the talk show rounds, raising questions about climate science and telling people not to worry. Polling shows a conversion rate of 34%, which keeps the majority of voters behind coal and petroleum. We expect the power and transportation markets to hold up for another 10-15 years.

And renewables?

We've moved $7.2 billion into wind & solar so far. The great thing here is that once we have the infrastructure in place, we'll be able to sell electricity at current rates without paying anything for fuel. This will increase energy profits around 614%.

What about extraction profits, Mike?

We're investing $20 billion to expand plastic production. That's enough to consume 95% of our petroleum stream by the time renewables kick in and drive a 40% expansion in extraction by 2050.

That's great, everyone. Thanks for the good work. I have lunch with the Senator at two, so let's call it a day.

There are certainly corporations that put the welfare of their employees and customers ahead of profits. But they are the exception. A corporation's legal responsibility is to maximize profits for its stockholders. If its goal was to lose money, it would be organized as a non-profit. We really shouldn't be surprised that oil, gas and coal companies would engage in media campaigns and lobbying to maximize the sale of fossil fuels for plastics. That's what they do. The old adage, *caveat emptor* (buyer beware), applies to plastics as much as it does to climate change, as much as it does to used cars. It's our responsibility to beware of greenwashing and to make personal choices that assure the health, safety and welfare of ourselves, our communities, our children and future generations.

There's no long-term solution to plastics. We can't manufacture our way out of this crisis. We can't control the industry or its regulators, but we don't have to wallow in plastics. If enough of us choose to, perhaps we can *deconsume* our way out of this crisis. In the meanwhile, you can reduce your personal exposure to plastics and their chemical additives while reducing your contribution to the plastic pandemic.

This Field Guide will help.

MAKING IT

These days, plastic is pretty much everywhere. Plastic[57] is strong, light weight and cheap so it's used in lots of great places. But it's also ending up in lots of places that aren't so great – like our forests, rivers and oceans. In our food, air and drinking water. In our bodies.

We've been making plastic-like substances for a really long time. People were tapping rubber trees and using the sap to make rubber balls and figurines nearly three millennia ago (1600 BCE).[58] In the 1800s we started modifying natural, plastic-like substances to make new things. For instance, we heated that rubber tree sap to make vulcanized rubber tires.[59] In the 1900s, we started making totally synthetic plastics things that had nothing to do with nature – things like neoprene, vinyl records, polyethylene, nylons, polyester, and teflon.[60]

The first big plastic boom came with World War II. War contracts expanded the fledgling plastics industry into giant manufacturers of military containers and packaging, goggles, helmet liners, parachutes, warplane parts and even atom bombs. A 1941 Harper's Magazine article said, "Because of the war, plastics have been turned to new uses and the adaptability of plastics demonstrated all over again."[61] And they were right! After the war, the plastics industry turned all those military things to household uses giving birth to polyester clothing, Tupperware®, Baggies®, Plexiglass®, and so much more.

The latest plastics boom started in the 2000s when the reality of the climate crisis created a surge in renewable energy solutions.

This loss of traditional revenues forced the petroleum industry to redirect their fossil fuels into plastics.[62] That's right. Plastic is made out of fossil fuels. The same fossil fuels that run our cars, heat our homes, and drive our power plants. The same fossil fuels causing global warming and all our climate problems. We're talking about crude oil, natural gas and even coal, being turned into plastics.

It's hard to imagine all that sludgy, dirty, smelly stuff being turned into nice, shiny plastic. But that's what happens. All plastic-like substances are made out of *polymers*[63], which are very long macromolecules composed of repeating chains of smaller molecules called *monomers*[64]. There are lots of natural things that are polymers. The sap from that rubber tree is a polymer. The proteins in our food are long polymer chains of amino acids. Our DNA is long, polymer chains of nucleotides. And plastics are long polymer chains of hydrocarbons – i.e. fossil fuels.

It says something about the ingenuity of the fossil fuel industry that they can take all the nasty stuff being mined-drilled-fracked out of the earth, take it apart at the molecular level, reassemble it into new substances that never before existed – then turn it into beautiful, shiny plastics. *Wow!*

The path from fossil fuel to plastic isn't hard to understand. Take crude oil, for example. When crude oil goes to a refinery, it's sorted into parts that have different weights. The refineries call these different parts *fractions*. Different fractions get used for different things. For instance, the heaviest fraction is used to make tar and asphalt roads. A less heavy fraction is use to make lubricants and motor oils. Increasingly lighter fractions make diesel, kerosene and jet fuel. The next lighter fraction is called *naphtha* and it is used to make synthetic chemicals and plastics. This naphtha fraction moves from the refinery to a cracking plant where it is *cracked*[65] apart into monomers, which are then combined into those long polymer chains that eventually get molded into plastic products.[66]

TRIVIA: The next lighter fraction after naphtha is used to make gasoline for our trucks and cars.

A refinery uses temperature to separate a fossil fuel into fractions. Think about how the steam from your hot shower hits the cold shower door and condenses, letting you draw pictures with your finger. Or how the inside of your car windshield frosts over on a cold morning from everyone's warm breath condensing on the cold windshield. Or maybe you've heard of a crazy uncle running a still that heats up mashed corn, runs the resulting vapor through some copper tubing to cool it to just the right temperature, and condenses it into moonshine alcohol.[67] A fossil fuel refinery works just like that. Different parts of a fossil fuel condense at different temperatures. So the refinery vaporizes the crude oil, runs the vapor through a distilling tower (a fancy version of that uncle's moonshine still), then pulls off the different fractions as they condense at different temperatures. Imagine if you could heat your shower door so it was hot on one end and cool on the other, then fill your shower with vaporized crude oil. You'd have gasoline condensing on the cool end of your shower door and diesel fuel condensing on the hot end.

A cracking plant uses chemicals to break a fraction down into monomers. If you've ever accidentally mixed household bleach with vinegar, you quickly learned that these substances react together chemically to become chlorine gas.[68] Or perhaps you've used a paint stripper to remove old paint from your furniture. These are examples of household chemical reactions that break something down into something else. That's basically how a cracking plant works. It adds chemicals to create a reaction that breaks naphtha down into monomers – a process called *catalysis*.[69] The monomers are then combined with more chemicals to knit them into long polymer chains – a process efficiently called *polymerization*.[70]

WARNING: Household bleach reacts with vinegar to release chlorine gas which can cause skin pain, redness, blisters; blurred vision, burning sensations in the nose, throat, lungs, eyes; chest tightness, difficulty breathing, respiratory failure.

The other things that get made out of cracked naphtha are synthetic chemicals. Thousands and thousands of synthetic chemicals. These are combined with the polymers to make them shiny, or waterproof, or to keep them from bursting into flames – or make them flexible, rigid, sticky, smooth, red, green or hundreds of other properties that make different plastics different.

So, combine lots of synthetic chemicals into your polymers and you finally have plastic resins, ready to be chopped into tiny beads and shipped to factories around the world where they're melted and molded into consumer products.

TRIVIA: The tiny beads of plastic resin that are shipped to factories around the world are called "nurdles". Some claim that the word was created by the writers of Rowan & Martin's Laugh-In TV show, along with farkel and bippy. The American Dental Association refers to the correct amount of toothpaste to put on your toothbrush as a nurdle. How this silly word became a not-so-silly environmental threat is unknown.

That's how plastics are made. Fossil fuels are refined into fractions that are cracked into monomers, that are polymerized into polymers, then compounded into resins. Poly is the key word here. Think polyester clothes, polyethylene sheeting, PVC pipe (polyvinyl chloride). The world of plastic is all about polymers and the thousands of synthetic chemicals that are part of the fossil-fuel-to-plastic pathway.

Take a moment, again, to wonder at the brilliance of the fossil fuel industry to be able to create something entirely new – substances that have never before existed. Then take a moment to wonder if there's a downside. We know that burning fossil fuels do lots of bad things to the environment. Is turning them into plastic any better? The Climate Reality Project says, "Cracker plants are the fossil fuel industry's latest attempt to lock us into a dirty energy economy, bringing extensive and expensive infrastructure that fills our air and water with toxic chemicals while contributing to the climate crisis and slowing the transition to renewables."[71]

We also need to wonder how nature will ever learn to cope with all these new synthetic substances. They're really no different than an alien substance crashing to earth on a meteor. Our planet's ecosystems have no idea how to decompose these alien substances. They'll just sit there for centuries – both the clearly named forever chemicals and the plastics themselves. This sits squarely on our shoulders.

We created plastics and we can uncreate them.

CODING IT

There are hundreds of plastic polymers and thousands of chemical additives. But understanding this complicated chemical world is not that hard. All plastics are grouped into just seven broad categories. Do you need to know this? No, but it's kind of interesting and, at the end of the day, this is really the easiest way you're going to know which plastics are annoying and which might kill you.

In 1988, the Society of the Plastics Industry created a set of symbols to help industry keep track of which plastic resins were where. These Resin Identification Codes (RIC) are tiny numbers inside a triangle. They appear on the bottom of most plastic products, although being very small and the same color as the plastic make them hard to find. Here's the codes:

Resin Identification Codes

----

1 – PET (PolyEthylene Terephthalate)
2 – HDPE (High-Density PolyEthylene)
3 – PVC (PolyVinyl Chloride)
4 – LDPE (Low-Density PolyEthylene)
5 – PP (PolyPropylene)
6 – PS (PolyStyrene)
7 – O (Other plastics)

I know. You're thinking, "Hey, those are the recycling codes." Well, the plastic industry says they're not, because "the presence of a RIC symbol on an item does not necessarily indicate that it is recyclable" and the industry certainly doesn't want to give you a legally-binding impression that plastics are recyclable.[72] Confusing? You bet! The RIC symbols were originally three arrows chasing each other in a triangle shape. Since the chasing-arrows have become so closely identified with recycling, the plastic industry has now changed the RIC codes to use a solid triangle. So, the numbers 1-7 inside a solid triangle are resin codes. The same numbers 1-7 inside a chasing-arrow triangle are recycling codes. *Geez!* We're just going to call them "codes" and talk about which are recyclable later.

So there are seven codes, but we can make this even simpler. First, notice that code names (PET, HDPE, LDPE, etc.) are just the initials of the chemical names. So nothing magic about the names.

Next, notice that codes #1, #2 and #4 are all polyethylenes, so we can just call all of these PE. The PE groups are used mostly for packaging and containers. The difference between PE groups is that one is high-density PE, one is low-density PE and one is what – normal density? Sounds like gym classes. Normal density (PET #1) is used for things like water bottles. They hold their shape but are easy to smush. High density (HDPE #2) is used for things like shampoo and detergent bottles that hold their shape and are hard to smush. And low density (LDPE #4) is stuff like those nasty grocery bags and cling wrap. No shape. Super smushable. So grab that plastic and squeeze. If you feel like superman, it's normal PE (PET #1). If it fights back, it's high-density PE (HDPE #2). And if it just goes limp in your fists-of-steel, then it's low-density PE (LDPE #4).

> *Trivia: Did you know that plastic brittleness is controlled by chemical additives? And what happens when something is too brittle? It breaks. Built-in obsolescence, anyone?*

Let's talk about code #5 next because it's used for a lot of the same things as high-density PE (HDPE #2). Code five is PolyPropylene (PP #5), which is more rigid and heat resistant than HDPE #2.[73]

It's also more resistant to acids. So it's used for hard plastic toys and plastic car parts, orange juice and ketchup bottles.

TRIVIA: With enough chemicals and energy all hydrocarbons (fossil fuels) can be converted to any chemical monomer, but cheap is king and less conversion is more profitable. So most PE comes from crude oil and most PP comes from natural gas. The paths-to-plastic are just shorter for crude oil's ethane to polyethylene (PE) and natural gas' propane to polypropylene (PP).

We jumped over code #3 which is PolyVinyl Chloride (PVC #3). PVC gets used for things like building products, garden hoses, and those metallized plastic potato chip bags. Oh, and vinyl records. PVC is 40% ethylene and 60% salt.[74] Yep, salt (sodium chloride) puts the chloride in polyvinyl chloride. Salt makes PVC more water-resistant and fireproof than the PE products.

Code #6 is PolyStyrene (PS #6). Remember those polystyrene food containers we used to get at the fast food places? It should tell you how nasty PS #6 is when even the fast food industry is getting rid of it. There's still a few restaurants that put your leftovers in PS because it's so cheap. Stop eating there. Really. Where you're still likely to see PS #6 today is the foam peanuts used to fill shipping packages. But even there, more and more companies are replacing PS #6 with bubble wrap and those air-filled LDPE #4 pillows.

TRIVIA: Styrofoam® is a trademark owned by Dupont. All Styrofoam® is foamed polystyrene, but not all foamed polystyrene is Styrofoam®. Think generic vs. brand name. Calling all polystyrene food containers Styrofoam® is like calling all sodas Coke®. Not right, but we're going to do it anyway.

Code #7 is Other (O #7), the catchall for all the plastics that don't meet the requirements for any other code. For instance, bioplastic lands here because it's not made out of fossil fuels. This is also where you find plastic products that contain a blend of other codes. For instance, CDs/DVDs have a plastic base layer, metallized middle layer, then an acrylic lacquer top layer. Or glass lenses which may be one type of plastic with a different scratch-resistant or UV-resistant polymer coating, or both.

Let's look at those codes again.

Code #1 – PET (PolyEthylene Terephthalate): One of the PEs. Short-term food packaging and polyester fabrics. Cheap and designed to fail. Soda and water bottles, cups, food jars, trays and clamshells, fast fashion clothing.

Code #2 – HDPE (High-Density PolyEthylene): Another one of the PEs. Longer-term storage. Usually opaque in color. Chemical containers, milk jugs, margarine tubs, detergent and shampoo bottles, reusable plastic shopping bags, toys, cereal box liners.

Code #3 – PVC (PolyVinyl Chloride): More water-resistant and fireproof than the PEs. Think long-term storage and flammable liquids. Gas cans, cleaning supplies and automotive product bottles, garden hoses, potato chip bags, pool liners, shower curtains, inflatable mattresses, plumbing, window frames, vinyl flooring, vinyl siding, disposable gloves, tablecloths.

Code #4 – LDPE (Low-Density PolyEthylene): The other PE. Slash and trash. Bread bags, "paper" milk cartons, paper towel and bathroom tissue overwraps, trash bags, frozen food bags, grocery bags, six-pack rings, shrink wrap, pallet wrap.

Code #5 – PP (PolyPropylene): Heat resistant and resistant to acidic foods. Yogurt tubs, cups, orange juice bottles, ketchup bottles, straws, hangers, rugs, furniture, luggage, winter clothing insulation, insulated cups, medical items, amber-colored pill bottles, coffee cup lids.

Code #6 – PS (PolyStyrene): Soft, lightweight with good insulation properties. To-go food containers, hot cups, plates, egg cartons, shipping cushions, cartons, trays, packing peanuts, home insulation.

Code #7 – O (Other plastics): Polycarbonate, nylon, ABS, acrylic, disposable tableware, kitchen appliances, computers, shrink-wrap, upholstery, awnings, diapers, safety glasses, CDs/DVDs, headlight lenses, bioplastics, and all plastics containing BPA.

Bottom line: Knowing your plastic codes is the first step to understanding the plastics around you – knowing which ones you can live with and which ones you can't.

* * *

Action Items

Print a cool-looking plastic code chart and stick it on your fridge and bathroom mirror.

Turn containers over as you use them and look for the triangle with the code. What does your handy chart say about that code?

Do you have a science nerd in the family? Get a giant, detailed plastic code poster and hang it in their bedroom or office.

Check out the plastic charts at www.ClimateGKC.org/plastics/.

LIVING WITH IT

This book opened by saying, "Where's the plastic? It's everywhere." Pretty snappy, huh? Sadly, it wasn't a joke. Recent decades have been filled with news about chemicals leaching into our food and home materials off-gassing various volatile organic compounds (VOCs)[75] into our air – and almost all of that was because of plastics.[76] You're going to be shocked by how much of your home is plastic.

But it's not all bad news. There's simple, effective steps you can take to protect your home from this cornucopia of toxic chemicals. And the first step is to understand where the plastics are.

In Your Kitchen

Your kitchen is absolutely filled with plastic. Plastic dinner and glassware, cooking utensils, pots and pans, food packaging, dry food storage, refrigerator storage, cleaning supplies, dish rags, dish towels, shelf liner, drawer dividers, gadgets, appliances, the refrigerator, the floor ...

DINNER AND GLASSWARE

Let's start with your plates and glasses. And we're not talking about your picnic things. We're talking about the good stuff. Melamine[77] dinnerware looks and feels just like ceramic or stoneware, but it's plastic. Your might even have a nice glossy set of acrylic[78] dinnerware that looks like china. Face it. Any dinnerware that is lightweight and unbreakable is plastic. Almost all colored drinking glasses are plastic. Nice drinking glasses are often Tritan®, Acrylic® or SAN Blend® plastics. You might even have an expensive set of crystal-like glasses made our of polycarbonate, the same material used to make bullet-proof glass.

A recent ad for expensive stemware reads, "Pure glasses make a sophisticated showcase for artfully crafted wines. Both beautiful and extremely durable, they're made from Tritan® crystal that maintains its sparkling clarity through years of daily use and entertaining. Set on a long, slender stem, the coupe is a modern classic for serving Champagne and other sparkling beverages, as well as vintage cocktails." That sounds like a really nice piece of glass, but Tritan® is a plastic copolymer made of stuff like dimethyl terephthalate , cyclohexanedimethanol, and 2,2,4,4-tetramethyl-1,3-cyclobutanediol.[79]

Real glass, ceramic and stoneware products chip and scratch over time. Plastics do not. And you can't trademark glass or ceramic. If your kitchenware claims to be scratch-resistant, durable, unbreakable, or it's trademarked®, then it's plastic.

> *TIP: It's almost impossible to tell plastic from real glass and ceramic. If it's not heavy and doesn't shatter into a thousand pieces when you drop it, your dishware and glasses are plastic.*

COOKING UTENSILS

Cooking utensils fall more easily into the wood, metal or plastic categories. And then there's silicone utensils. They look like plastic but are better, right? Not really. Silicone[80] is a plastic polymer made by reacting the element silicon with fossil fuels to produce a monomer named siloxane.[81] This monomer is then compounded with the usual suspect group of chemical additives to create a wide range of silicone resins. Siloxane is mixed with rubber tree sap to create silicone rubber. All silicones are plastics with a little natural product mixed in. So is it better than plastic?

Maybe, but not much.

POTS AND PANS

Plastic handles and lids on pots and pans are pretty obvious, but did you know that many non-stick surfaces are plastic? Teflon®, for example, is a brand name for the polymer polytetrafluoroethylene.[82] It has the distinction of being one of the original sources of "forever chemicals" and has generally fallen out of use because of its toxicity.[83] Another type of non-stick surface uses a water-repellant coating of polystyrene plastic.[84]

FOOD PACKAGING

And then there's food packaging (sigh). Take a moment to consider all the plastic that comes into your home with the groceries. Processed foods are understandably packaged in plastic, but even raw fruits and vegetables are sealed into single-use plastic bags. We used to bring food home in glass bottles, cans, paper and cardboard boxes. Now, even durable foods like dried beans come in a plastic bag. And one layer of plastic is not enough. Think about those individually plastic-wrapped pastries, gathered in a stay-fresh plastic bag, then packaged in an outer plastic bag for marketing. Or those loafs of bread in an inner airtight plastic bag and an outer marketing bag. They're even shrink-wrapping individual baking potatoes in plastic now.

Did you know that cans are lined with plastic.[85] A popular science project is to suspend a soda can in some drain cleaner and watch the metal dissolve to reveal a plastic bag holding the soda.[86] And what about metallic potato chip bags? They're plastic films coated with a microscopic (.5 micron) layer of metal.[87]

Single-use plastics are by far the biggest cause of the plastic pandemic and a lot of single-use plastics are food-related. Take a few minutes and do a simple kitchen waste audit. Wait until your kitchen trash bin is full, them spread it out and see how much of it is plastic.

FOOD STORAGE

Our kitchens are filled with Tupperware®, Baggies®, Ziplock®, Glad®, Saran® and other plastic storage products. Storing in plastic increases the likelihood of chemicals leaching into your food. Microwaving plastics guarantee that chemicals are leaching into your food.[88] Kitchen canisters for storing bulk items like pasta, cereals, flours, etc., are usually plastic. Even nice, glass-like canisters are often Plexiglas® or Acrylic® plastic.

AND SO FORTH

Your kitchen cleaning supplies are mostly plastic – from the obvious plastic soap bottles to the less obvious plastic sponges and scrubbers. Your kitchen linens are almost certainly NOT made out of linen – they're probably a polyester/rayon blend. Then there's all the miscellaneous corkscrews, cutting boards, drawer dividers, potato peelers, corers, and hundreds of other kitchen tools that are plastic.

Your countertop kitchen appliances are mostly plastic. Even your countertop (Formica®, Corian®, quartz, faux granite) is made from plastic.

> *TRIVIA: Formica is made with Melamine resin, just like those fancy Melamine stoneware dishes.*

Bottom line: Your kitchen is the highest concentration of plastics in your home, has the largest impact on the environment, and poses the biggest threat to your health.

In Your Bathroom

Unless your bathroom fixtures are antiques – faucets, shower heads, towel bars – they're probably plastic sprayed with a metallic chrome finish. Your bathroom towels and other "linens" are most likely microfiber or some other polyester blend. Your toothbrushes, floss, combs and hair brushes are plastic.

The myriad stream of plastic bottles and dispensers that make up a modern bathroom are certainly no surprise. What may be surprising is the plastics INSIDE your products. Microbeads are small plastic particles (<1mm) that are commonly used in cosmetics, toothpaste and exfoliating products.[89] The accumulation of microbeads in public waters became such a problem that the United States banned their use in "rinse-off cosmetics" in 2017. But we still have plastic microbeads in lots of products that don't rinse off, like makeup and sunscreens.[90] And most manufacturer's just replaced the rinse-off microbeads with FDA-approved rinse-off metallized-plastic glitters.[91]

In Your Laundry

Unlike cosmetics, toothpaste and exfoliating products, laundry detergents are not required to list ingredients. A 2019 lab study of 300 laundry detergents not only found plastics in 40% of the products, it identified 520 plastics polymers that are on the EU's list of toxic substances (ECHA, Annex XV Restriction Report).[92]

In Your Living Spaces

You probably don't see a constant flow of single-use plastics or plastic bottles in the other rooms of your home, but they still contain a surprising amount of plastic. Your floors are probably plastic – plastic fiber carpeting, plastic laminate floorboards, vinyl or other plastic tiles. If you have hardwood floors, they're probably coated with a plastic acrylic or polyurethane finish. Your windows are probably curtained or draped with plastic fabrics. Your furniture is probably upholstered with plastic fabrics or faux (plastic) leather. Your walls and ceiling are probably painted with acrylic or synthetic latex paint (plastics).

Most of your furniture – tables, lamps, storage units – are going to have plastic components. ALL of your electronics – computers, printers, wi-fi, televisions, music systems – are plastic. And then there's your personal items. What about your telephone? Plastic. Purse or billfold? Plastic. Ink pens? Driver's License? Credit cards? All plastic. Do you have a weekly organizer for your vitamins? A blood-pressure machine? Do you wear contact lenses?

In Your Closets

Your bedroom comes encased with the full compliment of plastic paint, carpeting and window dressings. It also contains your bedding and wardrobe, which is probably your second largest collection of plastics after the kitchen – maybe your largest if you prefer clothing to cooking.

The average person's wardrobe is over 60% plastic.[93] The vast majority of clothing today is polyester, rayon, nylon and acrylic. These are plastics that follow the normal fossil-fuel-to-plastic pathway, then are extruded into fibers and spun into petrotextiles (fabrics). Today's fast fashion industry is totally dependent on plastic fabrics. And cheap, throw-away fashion is right up there with single-use food packaging as major causes of the plastic pandemic.

The problems caused by chemicals leaching from plastics into your food, also apply to chemicals leaching from your clothing. In 2011, hundreds of flight attendants began reporting that new polyester uniforms were causing "blistering rashes, swollen eyelids crusted with pus, hives, and in the most serious case, breathing problems and allergic reactions." One attendant, John, had such severe reactions that he had to be taken off the plane and to the ER multiple times.[94] The airline and flight attendants' unions commissioned testing and found "tributyl phosphate, lead, arsenic, cobalt, antimony, restricted disperse dyes known to cause allergic reactions, toluene, hexavalent chromium, and dimethyl fumarate" in the uniforms.

This played out across multiple airlines over the coming years. A 2018 study at Harvard University found that the number of flight attendants with multiple chemical sensitivity symptoms doubled after the introduction of new uniforms. One of the study's authors said, "Flight attendants are the canary in the coal mine because of the length and consistency of their exposure," noting that other people are being exposed to the same chemicals in their clothing but don't notice it because they don't wear it as much.[95]

A pediatric nurse practitioner in Grand Rapids, whose toddler son developed chronic eczema with infected sores on his hands and knees, spent months replacing soaps, cleaning products, lotions and creams before discovering her son's polyester athletic clothes were causing the problem.[96]

John, the flight attendant who went to the ER in 2011 and had been in perfect health before the introduction of new uniforms, spent years seeking treatment for ongoing symptoms and died in 2021 of "cardiopulmonary arrest, secondary asthma."

Bottom line: Plastic IS everywhere and all of these plastic polymers contain thousands of chemical additives which may leach into your food, your skin, or off-gas into your air. Helping you understand the risks and how to avoid them is the goal of this book.

Microplastics And Microfibers

Along with all the plastic you see around you, there's a lot of plastic that you can't see. Microplastics[97] are tiny fragments of plastic debris smaller than five millimeters (<0.20"). They're created over time as environmental conditions break larger plastic debris into smaller and smaller ones. Plastic never dissolves into its constituent parts, but plastic will break into smaller and smaller pieces until they become too small to see with the naked eye. With each break, new plastic surfaces are exposed and the cycle of chemicals leaching from the plastic starts again.

A 2024 study that examined 16 categories of meat, fish and vegetarian meat alternatives, found microplastics in all 16 categories of food and in 88% of the samples tested.[98]

Microplastic fibers (microfibers) are tiny fragments of clothing with a diameter less than 50 microns (<0.0195").[99] They're released when clothing is manufactured or laundered. An average family's load of laundry releases 700,000 plastic microfibers per cycle.[100] We now have an estimated 14 million tonnes of microfibers that have washed into the world's oceans and we're adding another 500,000 tonnes a year.[101]

The University of Toronto reports that over 90% of water samples taken from the Great Lakes since 2013 have unsafe levels of microfibers.[102]

In Your Air

Several new studies prove the ease with which microplastics are transferred from land to air and then sent on intercontinental and trans-oceanic journeys to remote parts of the world.[103] Microplastics are being collected in increasingly remote, uninhabited regions of the French Alps, Greenland's glaciers, Pyrenean Mountains and the Arctic Ocean. Microplastics have been collected from high altitude clouds over Japan's Mount Fuji and Mount Oyama.[104]

A 2020 study found that over 74 tonnes of atmospheric microplastics – the equivalent of around three million plastic water bottles – settle each year on rooftops and gardens in

Auckland, New Zealand.[105] A 2021 study found microplastics in 65% of lungs autopsied.[106] A 2022 study found microplastics in 77% of human blood samples analyzed.[107] A 2023 study found microplastics in 85% of live lung tissues collected during surgery.[108]

In Your Water

It's pretty obvious that microplastics and microfibers are in our water. Plastic rain is the new acid rain. A 2020 study tested rainfall over US protected areas and found 1000 tonnes of microplastics raining down on US national parks and forests each year.[109] A 2022 study found microplastics in beaches at a level of 950 particles per kilogram.[110] There's trillions of microfibers in our oceans.[111] The 2023 world Ocean Race[112] took samples from every corner of the world's most remote oceans and found microplastics in every sample.[113] But what about our drinking water? What's going into our bodies?

A 2017 study found microplastics in 87% of tap water sampled from over a dozen nations – and in 94% of US samples.[114] In 2018, the State University of New York tested 11 major brands of bottled water from nine different countries from across the Americas, Asia, Africa and Europe – and found double the level of microplastics previously found in tap water.[115] A 2024 study by Columbia University and Rutgers University has now found concentrations of microplastics in bottled water at levels 10 to 100 times higher than the previous study.[116]

A 2020 study by Trinity College Dublin (Ireland) found that babies fed formula from standard plastic infant feeding bottles were drinking up to 16,200,000 microplastic particles per liter – around 3,000,000 particles a day.[117] Another 2020 study by the Università Politecnica delle Marche (Ancona, Italy), found microplastics in 4% of placentas analyzed.[118] This same group found microplastics in 75% of human breast milk samples analyzed in 2022.[119]

> *TIP: They still make glass baby bottles. The nipple is going to be plastic, but it's still better that an all plastic bottle. And a silicone nipple probably healthier than other plastics.*

In Your Food

If you're eating food on and storing it in plastic, you're obviously going to have some microplastics break off onto your food. Even opening food packaging can result in small pieces of plastic on your food.[120] But what about microplastics in the actual food itself?

A 2018 study at the Incheon National University (South Korea) found microplastics in 90% of commercial table salts analyzed worldwide.[121] These are both environmental microplastics collected with the salt and additional microplastics introduced during processing. A 2022 study by the Central University of Ecuador and the University of Alicante (Spain) found microplastics in 100% of canned tuna samples analyzed.[122]

In 2022, the United Nations Food and Agriculture Organization published an extensive analysis of microplastics in food commodities.[123] They found that the average person consumes 58,451 microplastic particles per year from tap water; 1,605,394 microplastic particles from table salt; 1,168,000 microparticles particles from mussels (if you eat mussels); and 23,915 microplastic particles from refined sugar.

A University of Newcastle (Australia) study in 2019 found that the average person consumes five grams of plastic a week.[124] Reuters News immediately turned this into an infographic[125] showing that we consume a porcelain soup spoon full of shredded plastic each week, a half-filled rice bowl of plastic each month, a cereal bowl full of plastic every six months, and a heaping dinner plate full of plastic every year. They went on to show that we eat a standard life buoy every decade and two full-size mobile recycling bins in an average lifetime.

Bottom line: Plastics are all around us and inside us too. We constantly breath, drink and eat plastics. Microplastics, and even smaller *nanoplastics*[126], pass through our body membranes into our muscles and organs. While the plastic polymers themselves are painfully stable, the thousands of chemical additives are not. The impact of plastics displacing normal body tissue and effects of their chemical additives leaching into our bodies is just now being studied. We'll look at the latest health studies in the *Surviving It* chapter.

> *TRIVIA: Nano- is the new micro-. A micro-something is one-millionth the size of a meter – a size called micrometer or micron. A nano-something is one-BILLIONTH the size of a meter. That's really small. That's REALLY small. An average bacteria is over a thousand times larger than a nanoplastic. We know that plastics continue to break apart into smaller and smaller pieces of plastic. So why have we never found nanoplastics in our bodies before? Nobody thought to look.*

* * *

Talking Points

Silicone is not natural. It's a plastic polymer mixed with a little silicon (sand) or rubber.

Action Items

Play "I See Plastic." Can be played by one, two or more. Take turns saying, "I see plastic" and name a plastic item in the room. An item can only be named once. One point per item. High score wins.

Audit your kitchen trash bin to see how much of your trash is single-use plastics.

Audit your clothing closet or a drawer to see how much of your clothing is plastic.

Play "I See Plastic Pro." Same rules as above but you have to correctly name the plastic's resin code to collect your point.

TRASHING IT

There's so much plastic today that scientists are discussing the formation of a Plasticene Epoch, marked by the presence of plastics in Earth's rock layers. Plastics are such a significant component of the present-day rock cycle that they can be used as markers of the age and character of the sedimentary deposits in which they are buried. They've already identified a new type of metamorphic rock that is a mixture of melted plastic, beach sediment, basaltic lava fragments, and organic debris.[127]

All of the plastics around us are going to end up in the trash at some point. Single-use plastics step into your life, wave, then dive head first into your trash bin. *Slash and trash.* Even recycled products have an end-of-use date and eventually end up in the trash.

In 1950, the world produced two million tonnes of plastics. By 2019, we were producing over 459 million tonnes annually.[128] As of 2019, the world had produced around 9.5 billion tonnes of plastics.[129] And, as of 2020, the global mass of plastics exceeded the mass of all land and marine animals combined.[130]

Most of this plastic is still piled up somewhere.[131] Or buried. Or floating around.

Plastics Produced (1950-2015)

55% sent to landfill
30% still in use
8% was incinerated
7% was recycled

Packaging is by far the largest contributor to plastic waste.[132] It's not only more than 30% of all plastic waste, it's almost double the next sector.

Plastic Waste (million tonnes)
2019 by Industrial Sector

Packaging – 142.60 mt
Construction – 76.90 mt
Transportation – 62.17 mt
Consumer Products – 46.44 mt
Textiles – 43.88 mt
Electronics – 17.31 mt

Not surprisingly, the US produces the largest amount of plastic waste per person. A 2016 study found the US and United Kingdom to be the world's largest producers of plastic waste. Russia and the EU came in at around one-half US levels. And the world's most populous countries, India and China, came in at less than one-fifth.

Plastic Waste (million tonnes)
2016 by Country: Total, per Person

United States – 34.02 mt, 105.3 kg
United Kingdom – 6.47 mt, 98.66 kg
Russian Federation – 8.46 mt, 58.66 kg
European Union – 29.89 mt, 54.56 kg
India – 26.33 mt, 19.88 kg
China – 21.60 mt, 15.67 kg

By weight, most plastic waste is made of polypropylene (PP #5), followed closely by the Other (O #7) group of specialty plastics. High- and low-density polyethylene (HDPE #2 and LDPE #4) make up the middle range of plastic waste, and polyethylene (PET #1), polyvinyl chloride (PVC #3) and polystyrene (PS #6) fill out the bottom.[133]

Plastic Waste (million tonnes)
2019 by Resin Code

PP #5 – 62.00 mt
O #7 – 57.76 mt
HDPE #2 – 44.71 mt
LDPE #4 – 40.21 mt
PET #1 – 21.81 mt
PVC #3 – 21.20 mt
PS #6 – 15.17 mt

Despite the huge numbers, the plastics industry makes sure that those of us in the US don't see mountains of plastic waste. If we could see the waste, we'd stop making the waste. So, the US is the world's largest exporter of plastic waste to other countries. Plastic waste is exported in bales. The Association of Plastic Recyclers specify that bales should be a minimum of 30"x42"x48". California specifies an ideal bale size of 48"x30"x60" with a weight of 1000 lbs.

China used to import 45% of US plastic waste, but they banned the imports in 2017.[134] This diverted around 500 million tonnes of annual plastic waste to Kenya,[135] and drove Kenya to enact the world's harshest plastic bag ban – fines of $38,000 or up to four years in jail for anyone that was caught producing, selling, or using a plastic bag.[136] However, the legislation failed to stop the flow of waste, because it didn't address international corporations shipping preexisting wastes to Kenya.[137]

In 2019, the world agreed to amend the Basel Convention – an international treaty on the control of transboundary movements of hazardous wastes and their disposal – to includes plastics.[138] This treaty has been accepted by 191 countries.[139] The US is a party to the treaty but has not ratified it and opposes the plastics amendment. Although shipments of plastic waste from the United States are now technically "criminal traffic as soon as the ships get on the high seas," US exporters continue to ship plastic waste to other countries.[140]

Trash Vortexes

There's around 200 million tonnes of plastic floating in the ocean with another 11 million tonnes being added every year.[141] That's around 171 trillion pieces of plastic,[142] and 1.5 quadrillion microfibers.[143] About 80% of ocean plastics come from land sources and 20% comes from marine sources like nets, ropes, and boats.

Surface plastics gather in ocean vortexes, or *gyres*. Ocean gyres are large, circular ocean surface currents. The center of a gyre tends to be very calm, so the circular motion of the gyre's currents push plastics toward the calm center, where they become trapped. There are five subtropical ocean gyres across the world.[144]

Subtropical Ocean Gyres

North Atlantic Gyre
South Atlantic Gyre
Indian Ocean Gyre
North Pacific Gyre
South Pacific Gyre

The largest is the North Pacific Gyre which covers 20 million square kilometers. It contains the Great Pacific Garbage Patch (GPGP), which is made up of an Eastern Garbage Patch between Hawaii and California, a Western Garbage Patch located near Japan, and a convergence zone that acts as a highway for plastics to move between the East & West garbage patches.[145]

The GPGP covers about 1.6 million square kilometers. The East Garbage Patch is about twice the size of Texas or three times the size of France. Overall, the GPGP contains nearly 80,000 tonnes of plastics. It is more of a miso soup of microplastics than a

Gilligan's Island of large plastics, with around 1.8 trillion pieces (50 kilograms) of mostly microplastic per square kilometer.[146] The large, identifiable plastics tend to be bottles, bottle caps, wrappers, plastic bags, and nurdles; with the occasional plastic lighter, toothbrush, ink pen, baby bottle, cell phone or toilet lid originating from the US, Japan, South Korea, China and Taiwan.

Degradable, Biodegradable Or Compostable

All organic things decompose. They rot. They break down into natural substances like carbon dioxide and water vapor. They do this with the help of organisms like bacteria and fungi (microbes). Different microbes like to eat different things and by eating it, *decompose* it into natural substances. Microbes absorb some of the substances to help them grow, and leave the rest for plants to eat so that animals can eat the plants. Then, when the animal dies, microbes decompose it back into its natural substances and the circle of life continues. That's what the plastics industry has in mind when they talk about a "circular economy." But plastics don't work that way.

Plastics are certainly organic – they're made out of hydrocarbons just like we are – but their hydrocarbons are assembled in a way that is alien to nature. It's kind of like Doctor Frankenstein taking a bunch of natural body parts and assembling them into something unnatural. Plastic hydrocarbons are assembled into long-chain macromolecules that don't exist in nature. So when a microbe comes along to take a bite out of plastic, it just goes, "Ouch!" and moves on to decompose something natural. Doctor Frankenstein's monster could never die. Neither can plastics.

What a plastic DOES do is break apart. It breaks apart again and again until it's a microplastic. Then again and again until it is a nanoplastic. And probably again and again until it's a picoplastic (one-trillionth of a meter), which we haven't even thought to look for yet.

TRIVIA: Picoplastics! You read it here first.

Plastics don't break down. They just break apart into smaller and smaller microplastics and microfibers. So what are degradable, biodegradable or compostable plastics?

If a plastic is marked *degradable,* it's just saying that it breaks apart. That's silly. Everything is degradable. Plastics degrade into microplastics, which isn't helpful. This is breaking apart, not breaking down. When we talk about plastics degrading, we're talking about them breaking apart into microplastics.

A plastic marked *biodegradable* is better. It at least implies that something is breaking down into natural substances. That it's *decomposing,* which is good. The problem with biodegradable plastics is that they need ideal conditions to decompose. In landfills or floating around in the ocean, the decomposition is going to stop or go anaerobic which produces harmful methane gas.[147]

And then there's *compostable* plastics which are great if it's not greenwashing. *Compostable* means that something easily decomposes into natural substances through normal interaction with microbes. Unfortunately, most plastic products that are yelling "compostable" are saying that they'll decompose under ideal industrial conditions. They're not going to decompose in your backyard or the neighborhood compost pile. They're just going to sit there like any other plastic. Whenever you see "100% Compostable" on the front, check the back for the small legal disclosure that says something like "Commercially Compostable Only."

Recycling

Recycling has long been touted as the magic bullet to fix plastic waste – the solution enabling a sustainable circular economy. But the reality is that only 7% of plastics are recycled into new products. Only certain types of plastic can be recycled. And each recycle cycle lowers the quality and increases the impurities in the plastic. At best, plastics can be recycled 2-3 times before they're forced into the landfill.[148] Plus, the higher cost of recycled plastics make them of less value to manufacturers, other than as greenwashing.[149]

Over 80% of the clothing you donate to used-clothing stores ends up baled and shipped to clothing dumps in places like Chile.[150] Over 93% of the plastics you put in your curbside recycling bin end up baled and shipped to dumps in Kenya, Thailand, Malaysia and Vietnam.[151] Recycling plastic is practically impossible and the problem is getting worse.[152]

As the quality of recycled plastic decreases, the levels of toxic contaminants goes up. New studies find that even a single recycle can be a vector for spreading toxic chemicals.[153] Existing toxins accumulate[154] and the process of heating and remolding plastic causes chemical additives to break down into new toxins like PCBs and dioxins.[155] This means that recycled plastics are increasingly being banned from child, food-contact, and other high-risk plastic products.

This magic bullet has lost its sparkle and it's time to start thinking about what's next.

Energy Recovery

With the failure of recycling to address the growing mountains of plastic waste, we're hearing more and more talk about plastic energy recovery.[156] In other words, let's incinerate (i.e. burn) plastic waste to run power plants. After all, plastic is made out of fossil fuels, right? Right!

But oh, the irony. To even consider that renewable energy would force fossil fuels to be turned into plastics, which would then end up being burned for power instead of using renewable energy. Mind blown! Consider the delight of fossil fuel executives who used to just get paid for burning their fuels. With energy recovery, they get paid for converting their fuels to plastics, then paid again for burning the plastics. Now that's a circular economy they can get behind.

Also, consider that burning plastics produce more greenhouse gas emissions than burning the fossil fuels in the first place. And that burning plastics release a much more deadly cocktail of carcinogens and dioxins. Even state-of-the-art commercial incinerators being studied in the Netherlands, Spain, Lithuania and Czech Republic release levels of dioxins and PCBs that exceed EU legal limits.[157]

You may also hear discussion of recovering energy by chemically converting plastic waste into a diesel-like biofuel. Again, the irony. To redirect fossil fuels away from diesel to make plastics that we then convert back to diesel. There has to be a better way.

The interesting conclusion of a plastic energy recovery argument is that the only viable use for fossil fuel is as fuel. Which brings us to bioplastics.

Bioplastics

Bioplastics are made out of plants, which is good. Right? Of course, fossil fuels are also made out of plants – plants that died millennia ago and fossilized into coal, crude oil and natural gas. So following bioplastic's logic, fossil fuels are made out of plants, which is good. Right? Obviously not.

PolyLactic Acid (PLA) is the most popular bioplastic monomer.[158] It's made from fermented plant starch from plants like corn, cassava, sugarcane and sugar beet. After that, it's handled like any other plastic monomer – mixed with chemical additives to create a resin that can be melted and injected into molds to produce consumer products like disposable cutlery, cups, etc. PLA is a resin code O #7.

You're going to see a lot of hype that bioplastics are 100% compostable, plastic-free, safe, eco-friendly, etc. That's not technically a lie, but it's not the full story either. Plastic-free – as in it's made out modern plants instead of fossilized plants that died millennia ago. PLA is still a polymer just like petroleum-based plastics. And yes, 100% compostable – at industrial composting facilities. Under normal home or community composting conditions, bioplastics decompose at the same rate as petroleum-based plastics, taking hundreds to thousands of years to fully decompose.[159] A lot of today's bioplastic hype is greenwashing.

In the absence of an industrial composting site, bioplastics are going to be dropped in the trash and end up buried in landfills or washed out to sea where they act just like petroleum-based plastic and break apart into microplastics.[160] In the future, things may change. Scientists at the University of California, Berkeley, have demonstrated ways to potentially make compostable plastics that break down easily with just heat and water.[161]

Bottom line: Bioplastics hold promise for the future, but the current crop has a long way to go before they solve the plastics crisis.[162] It's really better to just refuse that straw and to stop buying disposable dishware.

Plastic Eating Microbes

Sounds like a b-rated movie where the alien microbes land in a sleepy country town. A car splashes through an odd puddle on a dry road. The driver finds nothing outside, but his tire is flat when he comes back around the car. Tire changed, he starts back down the road, only to crash as his steering wheel dissolves. An ambulance rushes the driver to the hospital, leaving a trail of microbes as the driver is gurneyed into surgery. Lights flicker, life supports fail, a nurse's yogurt cup explodes ... well, you get the idea.

We've already talked about how plastic polymers are alien to nature. There's simply no microbes that know how to eat plastics, so the plastics just keep piling up in our landfills and oceans. But that doesn't mean some microbe in the future won't develop a taste for long-chain macromolecule polymers. Evolution is an amazing thing and microbes are the fastest evolving life on earth. They reproduce every 10-15 minutes so their colonies can rapidly expand into billions of microscopic individuals. And when you have billions of individual microbes differentiating themselves every 10-15 minutes, you're going to eventually find one that goes, "Yum! Plastic!"

Microbiologists have been looking for microbes with an appetite for plastics. It's only a matter of time before some microbe is born with the right combination of molecular enzymes to turn plastic into dinner. And when it happens, the sheer volume of plastic in the environment is going to give that microbe's *binary fission* offspring a huge evolutionary edge.

There's been some success finding microbes that will eat plastics at high temperatures. But maintaining a high temperature makes the process expensive and produces greenhouse gases. In 2023, a team of Swiss scientists found microbes in alpine and arctic soils that eat plastics at room temperature (59°F).[163] They tested the microbes on a variety of plastics and found that 19 strains were able to degrade biodegradable films, but none made a dent in standard polyethylene film.[164] That's certainly not going to save the world, but it does show potential.

In 2001, a team of Japanese scientists found a bacterium (microbe) feeding on plastic bottles, but didn't recognize its significance or publish their findings.[165] Fifteen years later, with the plastic crisis in full swing, they finished their earlier research and published in the journal Science, demonstrating that their bacterium (*Ideonella sakaiensis*)[166] was able to decompose PET #1 plastic.[167] That's really big, but the rate of decomposition was too slow for commercial application.

Scientists are now genetically engineering bacteria to increase the rate at which they can decomposition plastics.[168] Hey, if we can't make a plastic that normal microbes can decompose, let's make a microbe the can decompose normal plastics!

If we can ignore paranoid thoughts of GMO'd plastic-flesh eating bacteria, the natural and accelerated evolution of microbes appears to be the most promising solution to the plastic waste crisis. Combine this with a compatible line of bioplastics that free us from our dependence on fossil fuels, and we may have a future.

Now, what do we do about all the toxic chemical additives used in plastics?

* * *

Talking Points

Plastic molecules are alien substances that natural microbes can't decompose.

Only 7% of all plastic gets recycled. The rest is baled and shipped to foreign countries.

At best, plastics can only be recycled 2-3 times.

A single load of US laundry releases up to 700,000 microfibers into the ocean.

Action Items

Setup separate bins for your rubbish, paper goods, and plastics.

Wash and reuse your glass jars for storage. Recycle what you don't need.

Wash and recycle your cans and jar lids for scrap metal. You can get paid for your cans.[169]

SURVIVING IT

Because plastics don't decompose, we don't have to worry about them dissolving inside our bodies. That's good news. Or is it? Scientists are just beginning to look at what happens when our bodies start filling up with plastic particles. Plastic particles may not react with our body, but they take up space and what happens when a river silts up or is dammed? What's going to happen to our bodies when an organ, blood vessel or nerve channel becomes partially blocked with plastics?

We've already talked about eating it, drinking it and breathing it. Lots of studies have already found microplastics in our lungs, gut, blood, breast milk and placentas. We know it's there. A study published in *Frontiers of Endocrinology* reports that microplastics smaller than <20 microns can penetrate organs, and microplastics smaller than <10 microns can penetrate cell membranes.[170] A study published in *The New England Journal of Medicine* reports finding microplastics in over 50% of carotid arterial plaque removed from patients treated for clogged arteries.[171] A study published in *Particle and Fibre Toxicology* reports that micro- and nanoplastics can reach the brain and "inhibit AChE activity, alter neurotransmitter levels, and change behavior in several species," a fancy way of saying that it impairs memory, thinking, language, judgment and other thought processes.[172]

Here's the latest. A 2024 study by the *University of Rhode Island* introduced microplastics into the water of mice over a three week period.[173] They observed that the mice began to exhibit behaviors similar to human dementia. Particularly the older mice. With dissection, they found that microplastics had *bioaccumulated* in every organ, including the brain, liver, kidney, gastrointestinal tract, heart, spleen and lungs. Brain tissue tests revealed that the levels of Glial Fibrillary Acidic Protein (GFAP) had dropped, a condition associated with neurodegenerative diseases like Alzheimer's and depression.

Keep in mind that plastics have been piling up in our environment for 50+ years, but the awareness of microplastics in our bodies just hit the scene about five years ago. This is cutting edge stuff. We haven't had time to do long-term impact studies on humans. We don't even know if the microplastic concentrations in our bodies are high enough yet to have long-term impacts. The mice studies are eye-opening, but we shouldn't start worrying yet. We just need to know the microplastics are there, their concentration in our bodies is increasing, and that we need to start watching it. It's kind of like blood pressure. We didn't know it was this high. It's not going to kill us yet. But maybe we should start thinking about changing our diet.

What we DO need to worry about – and by "do" we mean NOW – is worry about the thousands of chemical additives in plastics. We have more than enough studies to know that these chemical toxins are affecting our health and behavior today. Some of these chemical additives have been around long enough that we've already done the studies, banned the chemicals (e.g. PCB, BPA), and are now running new toxicity studies on their replacements (Phthalates, BPS and BPF).

When you swallow a piece of plastic, your stomach acid and intestines try to dissolve it. When you breath microplastics, your lungs do the same. All our internal defenses work to dissolve foreign bodies like plastics. Of course, they can't. Plastic polymers are just too strongly bonded together. Nothing in nature can dissolve them and that includes our human bodies. But the chemicals additives in plastics are not as tightly bonded. They can be dissolved from the surface of plastics. This is called leaching – probably because of the way leeches draw blood from the surface of a host. Okay, that was a gross metaphor. Regardless, our bodies "leach" chemical additives from the surface of plastics. And, when a plastic particle breaks into two particles, new surfaces are exposed and the chemical leaching process begins again. Every time a piece of plastic breaks into smaller pieces, new surfaces are exposed and a fresh batch of chemicals leach into our bodies.

A 2019 *Consumer Reports* study of plastic products,[174] published in *Environmental Science & Technology,*[175] analyzed 34 everyday plastic products representing all seven plastic codes. They detected over 1000 chemicals, 80% of which were unknown. They then ran a series of toxicity tests and found that 70% of the plastics had some form of toxicity. They couldn't tell which chemicals were causing the toxicity – since most of the chemicals were unknown – but they could tell that something in the plastic was causing a toxic reaction.

A 2021 study by the American Chemical Society published in *Environmental Science & Technology* identified 10,547 *known* chemicals used as plastic monomers, additives and processing aids.[176] These were all chemicals registered with the Chemical Abstracts Service (CAS Registry).[177] Sixty-one percent of these chemicals had hazard classifications. Of these, 37% were low hazard, 12% medium hazard, and 12% high hazard, with 22 chemicals classed as PBT and 35 classed as vPvB. That's not Peanut Butter Toast. It's Persistent, Bioaccumulative Toxicity (PBT) and very Persistent very Bioaccumulative (vPvB).

Toxicity Levels of Identifiable Plastic Chemicals

39% – Unknown
37% – Low concern
12% – Medium concern
12% – High concern

Chemicals in the "high concern" category included chemicals with known CMR and EDC. That's Carcinogenicity, Mutagenicity, or Reproductive toxicity (CMR) and Endocrine-Disrupting Chemicals (EDC). In other words, they give us cancer, mess with our reproductive organs, and disrupt our thyroid, pituitary and adrenal glands.

In 2023, the United Nations did a study of chemicals in plastics.[178] They found over 13,000 chemicals. Of these, 7000 had been tested for toxicity and, of those, 3200 (46%) had one or more known hazardous properties. What about the other 6000 untested chemicals? Presumably designer-chemicals created by petroleum companies to make their plastics more competitive, with chemical properties that are trade secrets. It's probably hard to study trade secrets for toxic side effects.

So how toxic are the chemicals in plastics? We obviously consume plastics every day and are still walking around. The FDA says that most of the known chemical additives aren't going to effect us at low levels. Unless you're an infant. However, they agree that some of the chemicals can build up over time. They're PBT, baby. Persistent, Bioaccumulative Toxicity. Are you going to drop over dead? Probably not. Will your life be shorter? Maybe. Is your health and quality of life going to be effected? Yes.

As you might imagine, there's lots of things added to plastics. There's antioxidants, biocides, colorants, fillers, flame retardants, impact modifiers, light stabilizers, nucleating agents, odor agents, plasticizers and more. And there's so much overlap and cross use among all these additives that it's easy to get lost in the chemical maze.

Keep in mind that chemists are modern magicians. They pull chemical rabbits out of hats all the time. Not too long ago, they would have been called alchemists – or sorcerers, or witches. The classic pursuit of alchemy was the creation of the philosopher's stone[179] that allowed the transmutation of base metals like lead or copper, into silver and gold. Definitely shades of Harry Potter.

Modern alchemists – I mean chemists – turn base substances like fossil fuels into magical plastic polymers. They've finally found their philosopher's stone and can readily transmute one modern substance into another. And they're transmuting faster than rabbits. The Chemical Abstracts Service (CAS Registry) has more than 204 million unique chemical substances registered and is adding 15,000 new ones every day. That's mind boggling.

Plastic additives are all magical substances. Each comes in many variants and each variant comes with different properties that have different uses. A single additive can have one property that hardens and another that plasticizes. Another additive can waterproof in one variant and be a flame retardant in another. Lucky for us, there are additives and uses that stand out. We get to paint in very broad swaths that ignore most of the variants and alternative uses. Here are the standout additive types.

Types of Additives

Hardeners
Plasticizers
Lubricants
Waterproofers
Flame Retardants

Within each of these additive types is a standout chemical group that has the biggest impact on you and your environment. Let's focus on those.

Plastic Hardeners

Bisphenols are a group of 40 or so chemicals that make plastics and epoxy resins hard. They also make plastics resistant to acids and oils. You find bisphenols in things like reusable water bottles, baby bottles, CDs/DVDs, thermal papers (receipts and faxes), medical equipment and some dental sealants.[180] They're also used to line food and beverage cans to protect the metal from acidic foods like tomatoes and sodas.

BPA is probably the best known bisphenol. The BP stands for BisPhenol and the A tells us the variant. BPA has been added to pretty much everything for decades but, in 2008, its health risks became big news.[181] Animal studies linked BPA to cancer, high blood pressure, type 2 diabetes and cardiovascular disease. There was also concern that BPA was an endocrine disruptor, a fancy way of saying that it mimicked estrogen, androgens and thyroid hormones, which could interfere with reproductive and nervous systems, metabolic and immune functions, and the development and growth of babies.[182]

The FDA has always maintained that BPA is safe, but adjusted their position in 2010 to express "some concern" the BPA might effect the brain, behavior, and prostate glands in fetuses, infants, and young children, banning BPA in infant bottles, sippy cups and baby-formula packaging. Canada declared BPA a toxic chemical in 2010. France banned BPA in all food-related packaging, containers and utensils in 2015.[183] The EU seriously reduced allowable limits of BPA in food-related products and banned it from infant and child products in 2018. The EU also banned BPA in thermal paper in 2020.[184] That's a lot of banning for something that is of just of "some concern".

"BPA-free" has become a big marketing slogan as manufacturers rush to replace BPA with other bisphenol variants, often BPS or BPF. But new research indicates that BPS and BPF have the same health risks as BPA.[185] The problem's not with the variant. It's with bisphenol.

Plastic Plasticizers

Plasticizers make plastics more plastic. What? Before modern plastics existed, plastic was a synonym for the word malleable, meaning "capable of being shaped or formed." For instance, people used to soak ox horn in water, making it plastic enough to be shaped into lamp shades. So, in modern parlance, a plasticizer makes something more easily shaped or formed, including things made of plastics.

PCBs (Polychlorinated biphenyls) are a highly carcinogenic plasticizer that the EPA banned in 1979.[186] This increased plastics' reliance on phthalates, a group of 2-3 dozen chemicals. Like most things chemical, phthalates are referred to by their initials. Their names usually being with the prefix "Di" and end with the word "Phthalate". So, if you see an ingredient abbreviation that begins with D and ends with P, you're probably looking at a phthalate.

Common Phthalates

DBP (Di-n-butyl phthalate)
BBP (Butyl benzyl phthalate)
DEHP (Di(2-ethylhexyl) phthalate)
DiNP (Diisononyl phthalate)
DPHP (Di(2-propylheptyl) phthalate)
DiDP (Diisodecyl phthalate)

There's lots of studies linking phthalates to lots of health issues. Phthalates can increase prenatal mortality, reduce growth and birth weight, cause skeletal, visceral, and external malformations. They can disrupt arginine and proline metabolism, resulting in obesity. They can trigger early menopause. Phthalates are linked to type II diabetes, insulin resistance, breast cancer, thyroid cancer, high blood pressure, asthma ... it's a really long list.[187] In simple terms, phthalates are toxic to reproduction, disrupt endocrine functions, and are skin sensitizing.[188]

The EU banned six phthalates (DiNP, DEHP, DBP, DiDP, DNOP, BBP) from toys in 1999. Japan banned DiNP and DEHP from toys and DEHP from food-handling gloves in 2001. The US limited the levels of DEHP, DBP, and BBP, especially in toys and childcare articles, in 2008. In 2020, the EU limited the levels of all phthalates in flooring, coated fabrics and paper, recreational gear, kid's swimming aids, mattresses, footwear, office supplies, clothing, related accessories, and all "textiles that come into contact with the skin."[189]

The greatest volume of phthalates are used in PVC #3, which is used in food processing equipment and fast food packaging.[190] A 2024 *Consumer Reports* study of 85 foods (67 grocery store foods and 18 fast foods) found phthalates in 99% of the foods studied.[191]

Many of the health issues mentioned above are associated with a western diet heavy in processed foods. Wouldn't it be ironic if it was the food packaging and not the food that was the problem?

Plastic Lubricants

The purpose of lubricants is to help a melted plastic resin flow smoothly into a mold and release easily from a molding machine. The fancy name for plastic lubricants are surfactants – **surf**ace-**act**ive **age**nt**s** – because they act on a plastic's surface.

Alkylphenols are the family of compounds mostly widely used as surfactants. Alkylphenols are all xenoestrogens[192] – chemical compounds that imitate estrogen – and can disrupt endocrine functions just like PCBs, BPA, and phthalates.

Nonylphenol[193] is the major alkylphenol in plastics. Nonylphenol has been linked to what is becoming our standard list of health issues. It's been shown to promote obesity.[194] It can induce chronic kidney failure.[195] It's associated with breast cancer.[196] It affects secretions in the human placenta that can result in pregnancy loss.[197] It's even been shown to affect insulin signaling in the liver.[198] That last one was in rats though, so maybe it doesn't count.

In 2021, the EU placed limits on the use of nonylphenol in all clothing and textile accessories.[199] The US EPA has requested that industry voluntarily phase out the use of nonylphenol and plans to do more risk assessment.[200]

Plastic Waterproofers

Can you imagine plastic that isn't waterproof? Deflecting water seems to be one of the few good things we can expect from plastic. Yet there are plastic polymers whose sole purpose is to absorb many times their weight in liquids. These superabsorbent polymers[201] or *slush powders* are widely used in products like disposable diapers. So plastic waterproofers are definitely a thing. Like the lubricants, plastic waterproofers are surfactants – they act upon the surface of plastics to make it, in this case, repel water.

> *TRIVIA: Superabsorbent polymers (SAPs) were used in tampons until the 1980s when they were banned due to links to Toxic Shock Syndrome.*

Per- and Polyfluoroalkyl substances (PFAS) are a massive group of synthetic chemicals used to make plastics waterproof. The EPA's toxicity database (DSSTOX) lists over 14,735 PFAS. The National Library of Medicine's database (PubChem) lists over six million.[202]

The original PFAS was marketed by DuPont under the trade name Teflon®. Yep, your original non-stick pan. This invulnerable substance was discovered in 1938, trademarked Teflon® in 1945, rebranded PFA in 1972,[203] and banned by 180 countries in 2019.[204] DuPont has settled 3,500 personal injury cases for $670,000,000 over PFAS leaked into the Ohio River.[205]

PFAS are classified as persistent organic pollutants.[206] Persistent is the nice way to put it. Most people just call them *forever chemicals*.[207] Really. Go to Wikipedia, search for *forever chemicals* and you'll be redirected to "Per- and polyfluoroalkyl substances".

We expect plastics to contain PFAS. But did you know that PFAS are used to waterproof things like paper and cardboard fast food packaging, pizza boxes, microwave popcorn, eye makeup, nail polish, dental floss, shampoo and water-resistant clothing? Did you know that the over-the-counter fabric waterproofing and stain-resistance products you're spraying on your clothes, rugs and upholstery are PFAS?

> *WARNING: PFAS are used to make natural fabrics waterproof and stain resistant and are used to coat paper and cardboard food packaging. Anything that's waterproof probably contains PFAS.*

PFAS are leaching into your skin from clothing.[208] PFAS are leaching into your food from packaging.[209] And there's lots of studies linking PFAS to cancers (testicular, kidney, liver, pancreatic),[210] reproductive problems,[211] immunity issues and decreased vaccine response,[212] thyroid disease,[213] high cholesterol,[214] high blood pressure,[215] heart disease,[216] ulcerative colitis,[217] obesity,[218] and the list goes on.

Lots of older PFAS have been banned. Mostly the long-chain PFAS like PFOS[219] and PFOA[220]. For instance, Teflon® was a PFOA. Newer PFAS have shorter-chain molecules which are said to be safer because they can be eliminated from the body more quickly. But even DuPont admits that their new short-chain PFAS called GenX causes cancer in lab animals.[221] And a 2019 study at *Auburn University* found that short-chains may pose more risks than long-chains.[222]

Plastic Flame Retardants

Can you imagine having your polyester clothing burst into flames or the foam padding in your sofa spontaneously combust? They're made out of fossil fuels, after all, and are highly flammable. Fortunately, US regulators have put strong laws into place requiring substantial amounts of flame retardant chemicals to be added to plastics in all electrical and electronic equipment, computers, television, toys, textiles, furniture foam, foam insulation, and so forth.[223] That's good, isn't it?

Human tissue levels of flame retardant chemicals in the US are three to ten times higher than levels in the EU where there is less regulation,[224] and residents of California have one of the highest US tissue levels because of the state's super-safe anti-flammability standards. California is now revising their own standards to reduce the amount of flame retardants required in consumer products.[225]

Plastic companies are only too aware of the visual impact of their products bursting into flames. And they make sure that this never happens by adding massive amounts of flame retardant to their plastics. Studies have found that plastic TV and computer enclosures are up to 33% flame retardant by weight.[226] The Yale School of the Environment reports that, "Industry has become so reliant on flame retardants that as much as a third of the weight of plastics used in airplanes comes from one type of PBDE flame retardant."[227] That's a lot of flame retardant.

The PBDE (PolyBrominated Diphenyl Ether) mentioned above is now found in soil, air and water, dust, and animal tissues. It's been found in livers, kidneys, adipose tissue (body fat), brains, breast milk and plasma, and has been shown to cross the placenta during pregnancy and cause problems with motor development, attention-related behavior and IQ performance.[228] In lab studies, PBDE causes hepatotoxicity (liver), kidney toxicity, gut toxicity, thyroid toxicity, embryotoxicity, reproductive toxicity, neurotoxicity, and immunotoxicity.[229] That's a lot of toxicity.

> *TRIVIA: There's three major groups of plastic flame retardants. Brominated Flame Retardants (BFRs), Chlorinated Flame Retardants (CFRs) and Phosphate Flame Retardants (PFRs).*

A lot of flame retardants are persistent organic pollutants (POPs), toxic synthetic chemicals that are dangerous to the environment and our health. They're PBT, baby. Persistent, bioaccumulative toxicity. And there's a noticeable pattern of toxic industrial behavior – replacing a banned POP with a new one that is subsequently banned and replaced by a newer, new one that is banned and replaced with a newer, newer, new one, to infinity and beyond.

The UN Environment Program reports[230] that Mirex, one of the first chemicals used as a flame retardant, was listed as a persistent organic pollutant (POP) in 2004 and replaced by chlorinated chemicals (PCB, PeCB). These flame retardants were listed as POPs in 2004 and 2009, and replaced by brominated chemicals (PBDEs, decaBDE). They were listed in 2009 and 2017, followed by chlorinated paraffins (SCCP) listed in 2017, then PFOS/PFOA listed in 2019, PFHxS listed in 2022 and now phosphate chemicals (TDCP, TCEP, TnBP) which are ... wait for it ... in the CDC Toxic Substances Portal with the following notes.[231]

"A number of health effects have been observed in laboratory animals ingesting phosphate ester flame retardants: brain and kidney lesions and decreased fertility caused by tris(2-chloroethyl) phosphate (TCEP); urinary bladder lesions caused by tributyl phosphate (TnBP); liver lesions caused by tributoxyethyl phosphate (TBEP); kidney and liver lesions caused by TDCP; and adrenal and ovarian lesions caused by tricresyl phosphate (TCP). ... Increases in cancer have been observed

in laboratory animals exposed to TCEP (kidney tumors), TnBP (urinary bladder and liver), and TDCP (liver, kidneys, testes, and adrenal gland tumors)."

Thankfully, the CDC Toxic Substances Portal assures us that:

"Limited data are available regarding health effects in humans. ... No studies have been conducted to examine the toxicity of phosphate ester flame retardants in children. ... No studies have been conducted to measure these chemicals in blood samples from groups of people representative of the U.S. general population. ... There is not enough information available to determine with certainty whether or not phosphate ester flame retardants produce cancer in humans."

Doesn't that make you feel better?

Well, don't high-five yet. There's another concern with plastic flame retardants. Dioxins. Better known in miliary parlance as Agent Orange, dioxins are considered the world's most toxic substances. Dioxins are formed during the production of brominated flame retardants (BFRs) and also when plastics with BFRs are recycled into new products.[232] Dioxins leach from plastics and are released when plastics are incinerated at temperatures below 1832°F.[233] Dioxins affect brain development, disrupt thyroid and immune system functions, and are associated with increased risk of multiple cancers, and immune system damage.[234]

The good news is that the next cycle of plastic flame retardants look promising. They're exploring polymeric flame retardants, meaning that the flame retardant is a plastic resin that can be tightly bonded with the other resins, instead of chemical additives that will leach out over time.[235]

Now we can high-five!

Bottom line: Plastics' chemical additives are modern alchemy. Magical compounds transmuted into 15,000 new substances each day. And, just like magic, they have a light side and a dark side.

May the force be with us all.

By The Group

It's a good idea to look at things from several angles. It not only let's us see things that we may have missed, but it helps us remember things we did see. We looked at the types of additives above. Let's look at the same stuff by chemical groups below.

Chemical Groups

Bisphenols (Hardeners)
Phthalates (Plasticizers)
Alkylphenols (Lubricants)
Polyfluoroalkyls (Waterproofers)
BFRs, CFRs, PFRs (Flame Retardants)

Bisphenols are used as plastic hardeners and to make plastics resistant to acids and oils. Bisphenols like PCB and BPA are banned because they're linked to obesity, cancer, and endocrine problems in fetuses and children. The latest studies suggest that the new bisphenols (BPF, BPS) may not be any better.

Phthalates are used to make plastics flexible. Phthalates are toxic to reproduction, endocrine disrupting, and skin sensitizing. They're linked to type II diabetes, insulin resistance, breast cancer, thyroid cancer, high blood pressure, asthma and more. Because of plastic packaging, they're found in 99% of all processed and fast foods.

Alkylphenols are used to make plastics flow smoothly into molds and release easily from the molding machine. Alkylphenols like nonylphenol promote obesity, induce kidney failure, are associated with breast cancer, pregnancy loss, and insulin problems.

Per- and Polyfluoroalkyls, a.k.a. PFAS, a.k.a *forever chemicals,* literally last forever. They have the same type of strong molecular bonds as plastic polymers. There's nothing in nature that can break them down. There's nothing in human bodies that can break them down, so they're bioaccumulative. They enter your body tissues and stay there. The list of health concerns include cancers (testicular, kidney, liver, pancreatic), reproductive problems, immunity issues and decreased vaccine response, thyroid disease, high cholesterol, high blood pressure, heart disease, ulcerative colitis, and obesity.

Brominated, Chlorinated and Phosphate Flame Retardants (BFRs, CFRs, PFRs) keep our plastics from bursting into flames. BFRs and CFRs are now listed as persistent organic pollutants and generally banned from child and food-related plastics. Human studies of PFRs haven't been completed at the time of this writing, but rats exposed to PFRs have developed lesions and tumors on their brains, kidneys, bladders, livers, adrenal glands and ovaries.

By The Codes

Here's some tips to protect yourself from the worst that plastics have to offer.

In general, DON'T HEAT ANY PLASTIC. Some will be marked "microwave safe," so *toast once and toss* but don't reheat. Heating plastics in a microwave or dishwasher dramatically increases the leaching of chemical additives into your food or product.

In general, DON'T REUSE PLASTICS accept for types HDPE #2, LDPE #4 and PP #5. All other plastics contain lots of chemical additives that will leach into your food or product over time.

PET #1 (or PETE)

Don't heat. Don't reuse. Heating will increase leaching of chemical additives into your food. PET #1 has porous surfaces that allow bacteria to build up. You don't want to reuse that. *Slash it and trash it.* That sucks big time for the environment, but at least you're still standing.

> *TIP: What if you just stop buying food in PET #1 plastic? You can turn it over and check the code at the grocer just as easy as you can at home. Not buying food in PET #1 plastic is good for you and good for the environment. If you don't buy it, you won't heat it, you won't reuse it and you're sure your food hasn't leached chemicals from it. And you're not adding it to the landfill.*

HDPE #2

Don't heat. ***Reuse Okay.*** Used for cleaners, shampoos, milk and yogurt, so there's really no reason to heat it. And there's less chemical leaching, so reuse and extended storage is less of an issue.

PVC #3

Don't heat. Don't reuse. Heating will increase leaching of chemical additives into your food. PVC #3 contains lots of phthalates and other chemical additives that can leach out over time.

LDPE #4

Don't heat. ***Reuse Okay.*** Used for bread and shopping bags, so there's really no reason to heat it. And there's less chemical leaching, so reuse and extended storage is less of an issue.

PP #5

Don't heat. ***Reuse Okay.*** Used for some squeeze bottles, medicines, and drinking straws,[236] so there's really no reason to heat it. And there's less chemical leaching, so reuse and extended storage is less of an issue.

TIP: Americans consume 390,000,000 straws a day. Most of us are old enough to not need a sippy cup. Is there really any reason to use a straw?

PS #6

Avoid. Leaches styrene and a wide range of other chemical additives. Polystyrene is being phased out of common use.

O #7

Be aware. These specialty plastics include lots of different resins that use lots of different chemical additives. Electronics are over 33% flame retardant by weight. All plastics containing BPA are here. So are contacts and eyeglass lenses.

Speaking Of Your Eyes

Do you wear soft contacts? They're obviously plastic, but did you know that they're mostly made out of PFAS?[237] That's *forever chemicals* leaching into your eyes.

A 2020 study published in *Environment International* found that the odds of visual impairments (near-sightedness, far-sightedness, astigmatism) increased in proportion to blood levels of PFAS and that vitreous disorders (cataracts, retinopathy, glaucoma, macular degeneration) were adversely associated with long-chain PFAS exposure.[238]

Back To Melamine

Remember that Melamine dishware? Looks and feels just like ceramic or stoneware? Melamine is a hard, fire resistant compound, used primarily to produce melamine-formaldehyde resins, which are used to make Formica®, laminate flooring, dry erase boards and yes, melamine dinnerware.[239] Melamine is used as a flame retardant in textiles. It's used in pesticides and fertilizers. It can be foamed up and turned into cleaner sponges like Magic Eraser®.[240]

Melamine can also cause kidney stones, kidney failure and death.[241] Symptoms of melamine poisoning include irritability, blood in your urine, little or no urine, symptoms of kidney infection, and high blood pressure.[242]

The FDA has concluded that melamine is safe in food under normal circumstances. Except in baby formula, which is a total no-no. And apparently not in prolonged contact with acidic foods like orange juice or tomatoes. Or at high temperatures like baking or microwaving.[243] But the "rest" of the time it's safe. Really?

Setting aside the 2007 melamine contamination in pet food that killed over a thousand pets,[244] and the 2008 melamine contamination in infant formula that sickened 294,000 children,[245] – both of which were deliberate contaminations to increase the food's apparent protein content – a study by the *Taiwan Consumers' Foundation* found that 100% of sampled tableware released melamine when filled with hot food or drink,[246] and a study published in *JAMA Internal Medicine* found increased levels of melamine in the urine of people 4-6 hours after eating hot soup from a melamine bowl.[247]

Bottom line: Everyone agrees that you shouldn't heat food in melamine dishware. Everyone agrees that you shouldn't store acidic foods in melamine dishware. Studies make us wonder if we should even serve hot food in melamine dishware. Common sense makes us wonder why we even own melamine dishware.

Back To Your Closets

Let's take another peek in your closets. Like all plastics, petrotextiles contain lots of chemical additives. But not just the standard array of plasticizers and flame retardants. There are approximately 8,000 synthetic chemicals used to turn raw materials into textiles.[248]

A study by the Swedish Chemicals Agency to assess textile chemicals' risks to human health and the environment identified 2,400 substances, of which 350 were flagged as hazardous.[249] This included 76 regulated substances – thirteen environmentally hazardous substances, one endocrine disruptor, three allergens, and 58 carcinogenic, mutagenic and reprotoxic substances (CMRs).

Polyester accounts for 51% of the global fibre market.[250] Polyester is a PET #1 resin,[251] usually considered a single-use plastic with high potential for chemical leaching, particularly when heated. There must be some interesting alchemy to allow polyesters to be washed and worn for months. On the other hand, we've already discussed the illnesses and death associated with new polyester uniforms in the airline industry.

Environmental Disasters

As if all the microplastics and chemical additives we choose to surround ourselves with are not enough, we need to consider plastic-related environmental disasters. After all, the manufacture and transport of millions of tons of plastic resins (nurdles) and their chemical additives around the world is bound to have a few hiccups. Okay, maybe a lot.

For instance, in 2021, the container ship X-Press Pearl was delivering materials to plastic factories In India when it caught fire and sank. It spilled 1680 metric tonnes of LDPE #4 nurdles (70 billion nurdles) into the Indian Ocean off Sri Lanka, along with 9700 tonnes of epoxy resin, 1040 tonnes of caustic soda and 25 tonnes of nitric acid.[252] The Guardian reported the immediate deaths of 470 turtles, 46 dolphins and eight whales. Chemical

Engineering News reported that 600 sea turtles washed ashore over the following months, and an estimated 3000 were dead at sea.[253] Over 20,000 families can no longer fish the region.

The World Shipping Council reports that an average of 1382 containers of nurdles are lost at sea each year. That's approximately 26,687 metric tonnes or 1.1 trillion nurdles a year. For instance, accidents like the MSC Zoe spilling 11 tonnes of PS #6 nurdles and an unknown tonnage of HDPE #2 nurdles into the North Sea in 2019.[254] Or two container ships colliding in Durban, South Africa, and spreading 49 tonnes of HDPE #2 nurdles across 1243 miles (2000 km) of beaches in 2018.[255] Or 150 tonnes of PP #5 nurdles spilling into the waters off Hong Kong and washing across the China Sea to Macao in 2012.[256]

Plastics and their chemical additives are also shipped by train. In 2023, a Norfolk Southern freight train derailment in East Palestine, Ohio, spilled 129,000 gallons of vinyl chloride (used to produce PVC #3) along with 55,910 gallons of plastic monomers and 30,110 gallons of plastic lubricant.[257] The monomers caused irritation to the eyes and skin, sore throat and nausea, rashes and breathing difficulties. The vinyl chloride caused drowsiness, disorientation, numbness and tingling of the extremities, nausea, and is a known carcinogen.[258]

In 2023, The Department of Defense reported that 245 military bases were leaking PFAS *forever chemicals* into groundwater aquifers that are primary or secondary sources of drinking water for neighboring communities,[259] and the EPA issued the first-ever federal Clean Water Act enforcement action to stop Chemours Company from discharging PFAS into rivers in Ohio and West Virginia.[260]

In 2023, eleven previously undetected short-chain PFAS were found in North Carolina drinking waters.[261] The United Nations Human Rights Office of the High Commissioner issued a statement criticizing DuPont, Chemours and the US

Government.[262] Their statement said, "Even as DuPont and Chemours had information about the toxic impacts of PFAS on human health and drinking water, the companies continued to produce and discharge PFAS," and that "Health and environmental regulators in the United States have fallen short in their duty to protect against business-related human rights abuses, including providing the public, particularly affected communities in North Carolina, with the type and amount of information necessary to prevent harm and seek reparation."[263]

An EPA analysis, released in 2023, reports that US industry "disposed" of over 60 million pounds of PFAS between June 2018 and June 2023.[264] Disposal included incineration, landfilling, mixing with other fuels and burning it, discharging it into wastewater treatment systems, and injecting it underground. The report's authors noted that this is likely a dramatic undercount because PFAS waste is unregulated in the US. The report confirms that PFAS is associated with cancer, birth defects, developmental damage to infants, and impaired functioning of the liver, kidneys, and immune system.[265]

In Conclusion

The sky's falling! Asteroids are going to destroy the earth!

Just kidding. Things are worse than that.

No, just kidding again. But how DO we take action against this overwhelming sea of toxic chemical additives and ever growing cloud of microplastics? The minimum defense is to turn your plastics upside down and look at the codes. Know the limitations of each code and follow the rules. Seriously. Don't heat food in or reuse plastics that aren't designed for that. It's hard enough to believe a corporation when they say something good about their product. If they're saying, "WARNING: DO NOT MICROWAVE," you better believe it.

A better defense is to start removing plastic from your life. Yes, it's everywhere, but we don't have to let it pile up in our homes and closets. We don't have to wallow in it. *Living Without It* is really not that hard. We're going to talk about that next.

* * *

Talking Points

They're now finding microplastics in over 50% of plaques removed from clogged arteries.

Chemists invent 15,000 new substances every day.

Contact lens are mostly made of PFAS. (forever chemicals)

Action Items

Learn the pros and cons of each resin code.

Turn your plastics upside down. Check the codes at the store, not after you've brought them home.

LIVING WITHOUT IT

Finally! We get to talk about the things you can do to make it better.

You know that microplastics and chemical additives damage your health and longevity. You know that mountains of plastic waste are creating a toxic environmental wasteland. You understand that American and international corporations are going to choose profits over your health and well-being, and you're not surprised that US regulators perform risk assessments and seek voluntary phase-outs of toxic chemicals, while the UN writes letters accusing the US of human rights violations and the illegal export of wastes to poor countries.

You can let corporations drive a circular economy that feed their profits or you can take control of the circle. YOU. Not your neighbor. Not the community or the government. YOU can choose which products you buy and by doing so create a clean, healthy home for your family.

The plastics pandemic is upon us. You can't stop the corporate madness, but you can stop being a part of it. You can refuse to let the worst of the plastics pandemic in the door. You can reduce the onslaught of slash-and-trash single-use plastics streaming through your kitchen. You can eat foods that are free of plastic-leached toxins. You can be healthier, happier, save money and be part of a circular economy that works for you, not the corporations. There's a lot you can do!

We've already talked about *Living With It* and you know where the plastics are. Now we're going to talk about *Living Without It.*

Here's the eight steps:

Living Without It

Buy groceries that are not wrapped in plastic.
Store food in glass containers.
Use ceramic and glass dishware, metal & wood utensils.
Reuse your LDPE #4 bags for bag-type things.
Reuse your HDPE #2 bottles for cleaning supplies.
Buy wood furniture.
Eliminate redundant electronics.
Wear hemp, linen, cotton and wool.

In Your Kitchen

Your kitchen provides lots of opportunities to reduce your exposure to plastics and their chemical additives. The 3.3 billion tonne gorilla sitting in the room is single-use plastics.[266] Like any gorilla, it's big, it's definitely there, and we're kind of afraid to make it to leave. But it's really not that hard. A lot of your single-use plastics are from the constant stream of food-contact plastics coming through your kitchen. The chemicals leaching from these plastics are the biggest threat to your health and your biggest contribution to the plastic waste crisis.

Single-use plastics are 40% of all plastic waste. Stop and think about that. All of the health and environmental problems caused by plastics and 40% of it is for a one time use! Eliminating single-use plastics is not only the easiest thing you can do, it's the biggest. That's a win for everyone!

Processed foods are a big source of single-use plastics. Both at the grocery store and at the drive-through. Processed, freezer-to-microwave easy-meals are laced with microplastics and toxins. They're expensive and they're killing you. A lot of people don't know how to cook, but your health is worth learning a few culinary tricks. It's not hard to chop a salad or warm up a can of soup. Anyone can use a one-button rice maker. Anyone can throw a list of ingredients into a pressure cooker or crock pot, press the button and have a tasty meal a few hours later. Anyone can even drop a slab of protein into a skillet and take it out when it smells good. So let's start with the grocery store.

YOUR GROCERY STORE

Shopping for food can be easy and fun. You just need to learn some new behaviors. If you want, you can treat grocery shopping like a game, where you brave the plastic wilderness to make plastic-free food choices. Consider it the *Grocery Plastic Challenge.* Winners get to live a healthy life.

First, scope out your shopping options. All grocers have a bulk produce section. Find ones that also have a bulk dry goods section. Look for those small natural grocers that specialize in healthy, environmentally friendly products. See if there's a zero-waste grocer that specializes in package-free products. When possible, shop small and shop local. And don't forget the farmer's markets.

> *TIP: We love our bulk warehouse grocers, but boy do they wrap everything in plastics. Layers and layers of it. It's sad to say but, until they rethink their packaging, you'll find safer, healthier food elsewhere.*

Next, carry your reusable bags with you at all times. Keep them in the car or have a bag of bags by the door. Or both! Corporate stores love to sell you their branded reusable shopping bags, but keep in mind that they're plastic HDPE #2. Yes, that's better than a single-use grocery bag, but not if you buy a new reusable bag every time you go to the store. Keep your reusable bags where you'll use them. And find some nice cloth bags instead of plastic. They're healthier and can go in the laundry when needed.

> *TIP: Don't let the person at the farmer's market shove your fresh produce into a plastic bag.*

The easiest way to reduce single-use plastics and purchase healthier foods is to shop your grocer for healthy, plastic-free foods first. This lets you fill your cart with fresh produce, and food in glass jars and tin cans before you even look at the plastic alternatives. Most stores are laid out to encourage you to buy processed foods in plastic – they make higher profits on that stuff. You might need to make a couple passes through the store to shop in a healthy order, but the benefits to your health and the environment are worth it. Not to mention that the extra walking is good exercise.

Shopping Order

Bulk Produce
Bulk Dry Goods
Glass Jars
Tin Cans
Plastics

Shopping in this order will help reduce plastics and put healthier food on your plate.

Bulk Produce

Start in the bulk produce section. This is the healthiest food, has the least exposure to chemical leaching, and doesn't involve any single-use plastics. Buy what you need for the week.

> *TIP: Those big plastic bags of produce may look cheap, but they're not. The plastic bags promote ethylene gas and mold which makes the food age faster. Half of the food is going to rot before you eat it. Half may already be rotten when you get it home.*

Put the bulk produce in your reusable cloth bag. No, don't use the tear-off plastic bags. A head of lettuce and half dozen loose apples, potatoes or onions rolling around your cloth bag is not a problem. Remember that cloth bags are washable and that the tear-off single-use plastic bag will end up floating around the Giant Pacific Garbage Patch.

TRIVIA: Those tear-off produce bags are probably going to be a tempting green color and say biodegradable or compostable. Keep in mind that they require an industrial facility to break down and they'll probably end up in a landfill instead.

Bulk Dry Goods

Next, move to the bulk dry goods area. Again, healthy food with minimal exposure to chemical leaching. The dry goods are probably going to be in plastic bins, but you can't win them all. Again, don't use the grocer's plastic bags. Reuse your LDPE #4 bread bags or the grocer's bags you've kept from previous trips. Keep these reusable bags stashed in your bag of bags. Zero-waste grocers will weigh your glass jars when you enter their store and deduct the *tare weight* when you buy stuff in them.

TIP: You can also reuse your LDPE #4 bread bags in the produce section if you really can't stand having your vegetables touch each other in your cloth bag.

Glass Jars

Now, look for food in glass jars. These have a long shelf life, are healthy, super easy to prepare, and the glass jars can be washed and reused for food storage. They're better than tin cans. You can find things like fruits and vegetables, milk, juices, spreads, dressings, dips, sauces and condiments in glass jars.

Canned Goods

Next, look for food in tin cans. Again, they have a long shelf life, can be healthy and are super easy to prepare. Look for fruit and vegetables, sardines, tuna, salmon, soups, beans, chili ... even Spam® and Vienna sausages if your tastes run in that direction. Yes, the tin can may have a plastic epoxy lining but it's going to leach less chemicals into your food than a completely plastic container. And the tin can be recycled.

> *TIP: Ask your grocer's butcher or deli to wrap your meats in butcher paper, not plastic. If they say no, say, "I'll try somewhere else." You vote with your pocket book.*

Plastics

Last and most least, shop your plastics. You want to reduce the amount of food you buy in plastics. If you've done a good job above, there won't be many things left on your shopping list. Perhaps a few things like meats, plastic milk jugs, sauces and condiments you can't find in glass, plasticized boxes of milk or ice cream, bagged breads ... oh, and the really unhealthy stuff like chips and sodas. Try to avoid plastic films as much as possible. They leach the most chemicals into your food.

If you're taking the *Grocery Plastic Challenge*, your goal is to have as few plastic items as possible. Give yourself a point for every item you've bought without plastic packaging and subtract a point for every item in plastic packaging. Try to increase your score each trip.

Checking Out

Use the self-checkout or lay your loose produce on the checkout register's conveyor belt. If the attendant automatically tries to put your food in a plastic bag, just say, "No, thank you." Find a store that will sack your groceries in paper bags and boxes. These make great trash bin liners at home and can be recycled. And you have your cloth bags if they don't offer paper. Many grocers will give you a cash credit for using your own bags.

YOUR FOOD STORAGE

If you've shopped well, you're not coming home with a lot of food-contact plastics. For those few you did buy, remove the plastic and put the food in glass storage for the fridge or freezer. *Slash and trash.*

Never store food in plastic. Wash and reuse all the glass jars you get from the grocers and store food in them. Go through your cabinets and replace all your plastic food storage containers with glass. Good glass storage containers can easily go from freezer, to microwave, to oven, to dishwasher. Yes, they come with airtight plastic lids and silicon seals, but the lids should be BPA-free and the food won't touch the plastic. The lids come off to microwave or bake, so there won't be any heat-induced chemical leaching. And you can rinse the lids by hand instead of running them through the dishwasher.

You can freeze food in your reused HDPE #2 and LDPE #4 bags. These are both stable down to -148°F.

For storing your dry goods, find some large glass jars with airtight lids. Canning jars are perfect for this and come in a variety of sizes up to one gallon. They're cheaper than kitchen canisters. You can also wash and reuse large pickle, olive or other glass jars you bought at the grocers.

YOUR TRASH

If you follow the *Grocery Plastic Challenge* shopping order, you'll find that the amount of trash in your kitchen keeps getting smaller and smaller. Consider keeping separate trash bins for rubbish, paper goods and plastics. Your plastics bin should get smaller and smaller over time and is a great way to measure your progress. The paper goods bins will take your non-plastic packaging and can be recycled. Paper grocery bags make great liners for your plastics and paper trash bins.

Glass jars and tin cans can be rinsed and put in recycle bins. The rubbish bin goes to the landfill. If you compost, you may find that you don't even need a rubbish bin.

YOUR DISHWARE

Go through your cabinets and replace all your plastics with ceramic, stoneware, porcelain and glass dishware. Replacements are not expensive. You can pick up stoneware plates, bowls, glass glasses and stainless steel flatware at your local big-box store for a dollar apiece. Sets are even cheaper.

It can be hard to tell melamine and acrylic plastics from ceramic and glass, so read the labels and watch out for greenwashing. Fine Tritan® crystal is still plastic. If a porcelain dish is marked "BPA-free", it must be plastic.

Donate your plastic dishware to a second-hand store.

YOUR COOKWARE

The non-stick pan craze is where PFAS (Teflon®) got its start. Trash your non-stick pans, and donate your plastic cooking utensils to a second-hand store.

Use stainless steel and cast-iron pots, pans and skillets. These are safe, healthy, and cook well on all cooktops, including climate-friendly induction cooktops. Both stainless steel and cast-iron can be *seasoned* with oil to provide an effective non-stick cooking experience.

Use metal and glass bakeware in your oven. Use your ceramic, stoneware, and porcelain dishware in your microwave. Metal and wood cooking utensils are cheap, healthy and effective. Use silicone spatulas instead of nylon. Both are plastic, but the silicone contains less chemicals than nylon.

All of the above can be run through the dishwasher.

Ceramic non-stick and aluminum pans shouldn't be run through the dishwasher. The Ceramic pans are made by sandblasting the surface of an aluminum pan with a mixture of titanium and ceramic then coating it with silicon-dioxide.[267] The non-stick action happens because a thin layer of the surface comes off with the food, and running them through the dishwasher will take the surface off even quicker. Aluminum pans can react with steel pans in the dishwasher and cause corrosion and pitting. They also scratch easily, can give food a metallic flavor, may leach into your food, and are not induction-cooktop compatible.

YOUR APPLIANCES

Kitchen appliances are plastic, but at least they're not single-use plastics. Good appliances are made out of PP #5 which has less chemical additives. Some appliances may be PS #6 and ABS #7 plastics which have more chemical additives. Read the labels and avoid #6 and #7 when you can.

The best way to reduce kitchen appliance plastics is to have *fewer* kitchen appliances. Fewer appliances in the kitchen ends up being fewer plastics in the landfill. This doesn't mean that you have to sacrifice functionality. It just means that you eliminate appliance duplication. Do you really need a blender, a food processor and a smoothie maker? No, get one appliance that does all three. Do you really need a crockpot, a rice cooker and a pressure cooker? They make appliances that does all three. Do you really need a microwave, an airfryer, a convection oven, a toaster oven, and a toaster? No, get one appliance that does it all. Instead of having a dozen redundant plastic kitchen appliances, you can do everything with just a few appliances. You can save a lot of money and you can reduce your plastics too.

The next time you need an appliance, get a multi-use appliance and donate the ones it replaces to a second-hand store.

YOUR PREP GADGETS

Chopping food on a plastic cutting board is a perfect storm of microplastics. Replace your plastic cutting boards with wood. Same goes for plastic mixing bowls. Replace them with stainless steel.

Use your glass storage containers instead of plastic wrap. Metal pan lids can cover your stainless steel mixing bowls. Or you can just set one of your ceramic plates on top of the bowl.

WARNING: Aluminum can react with stainless steel in the presence of water – a process is called "galvanic corrosion." So it may be best to not to use an aluminum lid on a stainless steel bowl.

YOUR DRINKING WATER

A 2024 study finds that boiling tap water removes 80% of PET #1, PP #5 and PS #6 microplastics from your water.[268] When heated the microplastics combine with minerals in hard water and settle to the bottom. An easier solution that uses less power is a kitchen water filter. This can be a countertop or an under-the-sink filter. Just make sure the filter is rated one micron or smaller to remove microplastics. Search the internet for "1 micron filter."

TRIVIA: More expensive reverse-osmosis filters can remove nanoplastics, chemicals and anything else bigger than 0.0001 microns.

Bottom line: The steps outlined above can save you money, put healthier food on you plate, and will make your kitchen practically plastic-free.

Not In The Kitchen

For those times you want to eat out, go someplace you can sit down. You deserve it and there will be less packaging. If you have leftovers, ask the restaurant to provide a paper takeout container. If they say no, then say, "Oh, that's too bad. I like your food but I don't do plastic takeout." You vote with your pocket book.

Even fast food chains have seating areas, and you'll usually get better service across the counter than in the drive-through. Your fast food will be wrapped in plasticized paper or boxes, but you can say no to the plastic bag, no to the straw and no to the plastic lid. Sit down and enjoy your meal.

> *TIP: Wooden chopsticks are classy. Learn to use them. Traditionally for Asian, but they work with any food. Wash and reuse them. Keep some in the car and say no to the plastic cutlery, which is super soft, super PFAS, and will fill you up with microplastics.*

In Your Bathroom

Your bathroom is your second-largest stream of single-use plastics after your kitchen. It's filled with plastic bottles, plastic sundries like razors and combs, and lots of polymerized cosmetics.

YOUR HEALTH AND CLEANING SUPPLIES

Your bathroom is filled with bottles. Most will be PET #1 and HDPE #2. Maybe some PP #5. The PET #1 is single-use because of its high chemical levels. *Use it and lose it.* PP #5 doesn't recycle, so it's better for the environment to leave them at the store. The HDPE #2 bottles are recyclable and can be reused since the chemical levels are less. There's two strategies for reducing plastic bottles in the bathroom.

Always buy glass bottles when you can. If you have to buy plastic, turn the bottle over and select HDPE #2. You can at least recycle and reuse these bottles. PET #1 contains more chemicals and is single-use. PP #5 doesn't recycle.

The second strategy is to buy larger volumes. This lets you buy less plastic over time. For instance, buy the BIG bottle of mouthwash and use it to fill a smaller, reusable bottle for daily use. Same for the rubbing alcohol and hydrogen peroxide in your medicine cabinet. Buy it by the gallon and use it to fill smaller, reusable bottles for daily use. That puts less plastic in the landfill and saves you a lot of money. This strategy also works for cleaning supplies. Large bottles of vinegar, bleach and soap can be used to fill smaller, reusable bottles for daily use.

If you're DIY (do-it-yourself) inclined, you can reduce plastics even further and save boat loads of money. Replace all of your single-use plastic bottles of shampoo, body wash, window cleaner, counter spray, air freshener, furniture polish, scouring powder, floor cleaner, bathroom cleaner, etc. with just a few bulk purchases of things like white vinegar, Castile soap, rubbing alcohol and baking soda. Search the internet for "DIY cleaning recipes" or "DIY shampoo recipes" or "DIY body wash recipes" and mix your DIY cleaner into a glass or reused HDPE #2 bottle. This eliminates all of those small single-use bottles, removes a lot of unnecessary chemicals from your home, and saves you a surprising amount of money.

Another great way to reduce plastics is to buy bar soaps and bar shampoos wrapped in paper. Plastic bottles are ... well, plastic. And both body wash and shampoo probably contain plastic exfoliates and other plastic-related substances.

TIP: Buy powdered laundry detergent and dishwashing soap in cardboard boxes.

YOUR SUNDRIES

Get a metal razor handle that accepts a standard razor blade refill. The cartridges are going to be plastic, but at least your not tossing a plastic handle into the dump every time you need a new blade.

Next time you need a comb, hairbrush or toothbrush, shop for wood or bamboo instead of plastic. Wood combs don't create static in your hair the way plastic does. Bamboo toothbrushes still have nylon (O #7) bristles, but at least you're not tossing a plastic handle into the dump.

Use wood toothpicks or get a metal pick instead of single-use plastic floss picks. Search the internet for "metal floss pick."

YOUR COSMETICS AND MAKEUP

Cosmetics and makeup come with lots of plastic packaging and contain a surprising amount of plastic substances. Studies find microplastics in 87% of all cosmetics.[269] Modern makeup formulas use plastic resins – often copolymers of acrylic[270] and nylon[271] – mixed with various blends of UV light stabilizers, plasticizers (phthalates)[272] and waterproofers (PFAS)[273] to give makeup its pizzazz. Some nail polishes and eye shadows actually contain plastic glitter. The EU has banned glitter from rinse-off cosmetics starting 2027, from leave-on cosmetics starting 2029, and from makeup and nail cosmetics starting 2035.[274]

Try to buy makeup with as little plastic packaging as possible. Here again, buying a larger container of one product will save you money and put less plastic in the landfill over time. For instance, an eyeshadow palette with 1-2 shades will contain much less plastic packaging than a palette with a dozen or more shades.

Search the internet for "DIY cosmetics recipes" of "DIY makeup recipes" to find plastic-free solutions.

In Your Closets

Check the labels and you might find that you have more plastic in your closets and drawers than you do in the kitchen and bathroom. Polyesters, acrylics, nylons, rayons, lycras and spandex are all plastics. Most of today's fabrics are partially, if not completely plastic. Your clothes, bedding, towels, upholstery, carpets and curtains all probably have lots of plastic in them.

Natural fabrics like hemp, linen, organic cotton and wool are comfortable, healthy and kind to the environment. They're better than plastic. You don't need to rush out and replace everything you own, but next time you buy a fabric, look for a 100% natural fiber. Refuse to buy plastic, even a cotton/poly blend.

> *TIP: Unless you fall down, you don't need to wash clothing every time you wear it. Wait for it to fail the sniff test. Doing laundry less often and hanging it to dry makes your clothing last longer, saves you money and releases fewer plastic microfibers into the environment.*

In Your Living Spaces

You won't find a lot of single-use plastics in your living spaces. That's good! What you will find in you living rooms, offices, and bedrooms are lots of hard plastics with large concentrations of flame retardants. Flame retardants don't leach out as much as they shed small particles into the air and household dust.

YOUR FURNITURE

Polyester upholstery and plastic foam padding make sofas and chairs really flammable. Safety regulations compensate for this by requiring high levels of flame retardants in these products.

Furniture foam padding breaks apart and off-gasses as it ages, filling a room's air and dust with chemicals.

Plastic leather, often called pleather, is made out of PVC #3. This class of plastics has a long list of leachable chemicals.

Your best solution for chemical-free furniture is wood with cushions and pillows covered in natural fiber upholstery and padded with cotton or other natural filling.

YOUR MATTRESSES

Not surprisingly, foam mattresses are full of synthetic latex, vinyl, flame retardants and other chemicals like plasticizers (phthalates) and fluorocarbons.[275] They off-gas volatile organic compounds (VOCs) throughout their life. When you lay on one, the heat from your body will increase the off-gassing.

Adding extra layers of foam toppers to a bed probably doesn't help. It's just more plastic. A mattress case can help contain the off-gassing if the case is impermeable. But an impermeable mattress case is also going to be plastic with it's own set of chemicals and VOCs. Obviously, you've been sleeping on that foam mattress a long time and haven't died. But it does make you wonder about various respiratory problems and headaches that you might experience.

There are foam mattresses that take extra steps to reduce toxic chemicals and lower VOCs. It's difficult to tell how much of this is greenwashing. Search the internet for "chemical free mattresses." The natural alternative to sleeping on plastic is mattresses made out of cotton, wool, steel springs and natural latex (rubber tree sap).

YOUR ELECTRONICS

Your electronics are another group of products that could understandably burst into flames and are required by law to contain high levels of flame retardant chemicals. Your TV, laptop, tablet and phone's plastic cases can be up to 33% flame retardant by weight and contain a full share of plastic hardeners (bisphenols). All of these electronics contribute to toxins that accumulate in your home's dust and air. Dry mop around your TV and you'll come up with a handful of BFRs.

Nobody is going to stop watching TV or using their phone. But do you really need a TV in each room? Do you need both a tablet and a phone? Do you need multiple computers? The answer may be yes. For some, reducing the number of electronics can help reduce the level of plastic-related chemicals in your home's dust and air.

There's also the question of plastic waste. All electronics are resin code O #7 and can't be recycled. If you own three TVs, you're eventually putting three TVs in the landfill. If you just have one TV in say a "TV room," you'll only add one to the landfill.

YOUR TOYS

Toys are an important part of a child's development. They help the child develop motor and cognitive skills. Playing with toys that might leach chemicals known to cause neurological and behavioral issues is a bad idea. Even in the US environment of voluntary phase-outs and ongoing risk assessments, potentially toxic plastic additives are quickly banned from toys and other child-care products.

Before plastics, children progressed from wood blocks to tinker toys to metal erector sets. Plastics came along and protected children from the physical cuts, scratches and splinters of this previous generation of toys. But now we know that plastics brought their own subtle form of chemical injury.

Wood and metal toys are making a comeback. Tinker Toys® and Erector® sets are readily available again along with plenty of other wood and metal options.

TIP: If you're buying wood and metal toys to avoid plastics, don't forget to check the packaging. It seems counterintuitive, but your new wood or metal toys could come in a HDPE #2 or PP #5 plastic storage box.

Product Packaging

Another source of single-use plastics streaming into your home is product packaging. Over 30% of all plastic waste is packaging. *Slash and trash.* Think about those giant, clear, hard plastic packages that come on the smallest of items. Think about how much you struggle to get them open at home. Imagine what they're going to do to the environment. When you're shopping, try to purchase products that are packaged sustainably. And refuse the plastic bags at the checkout register. Your cloth shopping bags aren't just for grocery stores.

Don't forget the mail. For decades shippers have been padding packages with polystyrene panels and peanuts (PS #6). We now know that PS #6 is really toxic and shippers have started padding with LDPE #4 pillows. That's definitely progress, but we can do better.

The latest packing material is mycelium. That's the fancy word for mushrooms. Fungus. Some smart people figured out that they could create a custom-shaped mold and get mushrooms to fill all the open spaces with their roots. Remove the mold, dry it, and you have perfectly shaped padding to slip your product into for shipping.[276] You've probably already received packages with mushroom packaging. It looks like an off-white to brownish paper mache that fits snugly around the product. Put it in your compost bin. It easily composts in about 45 days.

Plastic in the mail is still an issue. Before you order something check the company's packaging policies. If you receive a shipment filled with polystyrene, let them know why you won't buy from them again. Let them know that there are better solutions than plastic packaging. Vote with your pocket book.

You're Not Alone

It's not just you. People around the world are feeling the health and environmental impacts of plastics.

In 2022, all 193 UN Member States endorsed a resolution to negotiate an international legally binding agreement to end plastic pollution.[277] The resolution established an Intergovernmental Negotiating Committee tasked with completing the draft agreement by November, 2024.[278]

Politics aside, the UN Environmental Programme (UNEP) has published a report showing how 80% of global plastic waste can be slashed by 2040.[279] The steps outlined in the report are solid and include the steps outlined in this book. They include the elimination of single-use plastic packaging, the promotion of reusable plastics, and the replacement of fossil-fuel plastics with biodegradable alternatives.[280]

Refuse, Reduce, Reuse

Living without plastic is really not that hard. You can refuse to buy new plastics, reduce the most toxic plastics around you and reuse the plastics you already own.

The best thing you can do for yourself and the planet is to stop the flow of single-use plastics through your kitchen and bathroom. Stop cooking, eating and bathing in plastic. The steps we've outline are reasonable, will stop the toxins leaching into your food, and will save you money. What's not to like?

The next thing to do is to get control of your wardrobe. Petrotextiles are a huge source of toxins and plastic waste. A healthy wardrobe makes you feel good and look good.

The final thing to do is to reduce the plastics in your living spaces. You're not going to eliminate everything, but cutting back on plastic furniture and electronics will make your air more breathable.

Doing all the things outlined in this book creates a circular economy that works for people like you, not corporations. Vote with your pocket book. The choice is yours.

* * *

Talking Points

Fruit and veggies packaged in plastic spoil quicker, because the plastic make it age faster.

There's just eight steps to living without plastics.

The order you shop in determines how much plastic you buy.

Action Items

Take the *Grocery Plastic Challenge*. Winners get to live a healthy life.

Replace all your plastic dishware with ceramic, stoneware, porcelain and glass.

Throw away your non-stick pans and donate your plastic cooking utensils to a second-hand store.

Own fewer kitchen appliances by using multi-purpose appliances.

Sit down when you eat out. Say no to fast food's plastic bag, straw, and plastic cup lid.

Buy glass bottles instead of plastic.

Buy by the gallon and pour your product into a reusable bottle for daily use.

Don't play with plastic toys.

Own fewer electronics.

Consider the packaging when you buy something.

Vote with your pocket book.

ABOUT THE AUTHOR

Bob Grove grew up among northern Missouri farmers where he developed a passion for the rights of people and for healthy environments in which those rights can thrive. He currently serves as President of the Climate Council (www.ClimateGKC.org) and Executive Producer of the Climate Hour (www.ClimateHour.net). Bob is a frequent speaker on a wide range of sustainability topics, both in person and through online workshops. He can be reached through Wordsinger Publishing (www.Wordsinger.com).

ENDNOTES & BIBLIOGRAPHY

Because internet content can change over time, we do not guarantee that the URLs provided here will continue to work.

---

[1]. Wikipedia contributors. (2024a, January 18). *Timeline of plastic development*. Wikipedia. Retrieved January 26, 2024, from https://en.wikipedia.org/wiki/Timeline_of_plastic_development

[2]. Wikipedia contributors. (2024, February 7). *Chemical element*. Wikipedia. Retrieved February 13, 2024, from https://en.wikipedia.org/wiki/Chemical_element

[3]. Wikipedia contributors. (2023, December 2). *Molecule*. Wikipedia. Retrieved February 13, 2024, from https://en.wikipedia.org/wiki/Molecule

[4]. Wikipedia contributors. (2024a, January 6). *Chemical compound*. Wikipedia. Retrieved February 13, 2024, from https://en.wikipedia.org/wiki/Chemical_compound

[5]. Wikipedia contributors. (2023a, October 22). *Monomer*. Wikipedia. Retrieved February 13, 2024, from https://en.wikipedia.org/wiki/Monomer

[6]. Wikipedia contributors. (2024c, February 8). *Polymer*. Wikipedia. Retrieved February 13, 2024, from https://en.wikipedia.org/wiki/Polymer

[7]. Wikipedia contributors. (2024c, February 12). *Periodic table*. Wikipedia. Retrieved February 12, 2024, from https://en.wikipedia.org/wiki/Periodic_table

[8]. Whitten, K. W., Gailey, K. D., & Davis, R. E. (1992). *General*

Chemistry. Saunders College Publishing.; Wikipedia contributors. (2024, January 25). *Covalent bond*. Wikipedia. Retrieved January 26, 2024, from https://en.wikipedia.org/wiki/Covalent_bond

[9]. Flowers, P. (2019c, February 14). *2.4 Chemical Formulas - Chemistry 2E | OpenStax*. Retrieved February 12, 2024, from https://openstax.org/books/chemistry-2e/pages/2-4-chemical-formulas

[10]. *Making plastics: From monomer to polymer*. (2016b, July 26). AIChE. Retrieved February 14, 2024, from https://www.aiche.org/resources/publications/cep/2015/september/making-plastics-monomer-polymer

[11]. Wiesinger, H., Wang, Z., & Hellweg, S. (2021). Deep Dive into Plastic Monomers, Additives, and Processing Aids. *Environmental Science & Technology, 55*(13), 9339–9351. https://doi.org/10.1021/acs.est.1c00976

[12]. Wikipedia contributors. (2024d, February 9). *Nurdle (bead)*. Wikipedia. Retrieved February 13, 2024, from https://en.wikipedia.org/wiki/Nurdle_(bead)

[13]. *Fractional distillation - Energy Education*. (n.d.-b). Retrieved February 13, 2024, from https://energyeducation.ca/encyclopedia/Fractional_distillation

[14]

. *Refining crude oil - the refining process - U.S. Energy Information Administration* (EIA). (n.d.). Retrieved February 12, 2024, from https://www.eia.gov/energyexplained/oil-and-petroleum-products/refining-crude-oil-the-refining-process.php

[15]. Wikipedia contributors. (2024, January 11). *Resin identification code*. Wikipedia. Retrieved February 14, 2024, from https://en.wikipedia.org/wiki/Resin_identification_code

[16]. Wikipedia contributors. (2024b, January 18). *Natural-gas processing*. Wikipedia. Retrieved February 14, 2024, from https://en.wikipedia.org/wiki/Natural-gas_processing

[17]. Tian, P., Wei, Y., Ye, M., & Liu, Z. (2015). Methanol to Olefins (MTO): From fundamentals to commercialization. *ACS Catalysis, 5*(3), 1922–1938. https://doi.org/10.1021/

acscatal.5b00007; Wikipedia contributors. (2024c, February 11). *Coal liquefaction*. Wikipedia. Retrieved February 14, 2024, from https://en.wikipedia.org/wiki/Coal_liquefaction

[18]. Rodríguez Vallejo, D. F., Guillén Gosálbez, G., & Chachuat, B. (2020). What Is the True Cost of Producing Propylene from Methanol? The Role of Externalities. *ACS Sustainable Chemistry & Engineering, 8*(8), 3072–3081. https://doi.org/10.1021/acssuschemeng.9b05516

[19]. Liu, R. (n.d.). *Black coal to white trash*. New Security Beat. Retrieved February 14, 2024, from https://www.newsecuritybeat.org/2020/02/black-coal-white-trash/

[20]. Center for International Environmental Law. (2023, March 17). *Plastic & Climate: The Hidden Costs of a Plastic Planet (May 2019) - Center for International Environmental Law*. Retrieved February 14, 2024, from https://www.ciel.org/reports/plastic-health-the-hidden-costs-of-a-plastic-planet-may-2019/

[21]. Bryce, E. (2021, January 18). How do we turn oil into plastic? *livescience.com*. Retrieved February 12, 2024, from https://www.livescience.com/how-oil-is-turned-into-plastic.html

[22]. Kelly, M. L. (2021, December 1). Why Americans buy so much stuff: A short history. *NPR*. Retrieved March 10, 2024, from https://www.npr.org/2021/11/29/1059861668/why-americans-buy-so-much-stuff; Cohen, L. (2004). *A Consumers' Republic: The Politics of Mass Consumption in Postwar America*. Retrieved March 10, 2024, from https://dash.harvard.edu/handle/1/4699747

[23]. *The Age of Plastic: From Parkesine to pollution | Science Museum*. (n.d.). Science Museum. Retrieved March 11, 2024, from https://www.sciencemuseum.org.uk/objects-and-stories/chemistry/age-plastic-parkesine-pollution

[24]. Wikipedia contributors. (2023, November 20). *War Production Board*. Wikipedia. Retrieved March 10, 2024, from https://en.wikipedia.org/wiki/War_Production_Board

[25]. Wikipedia contributors. (2024, March 7). *Military–industrial complex*. Wikipedia. Retrieved March 10, 2024, from https://en.wikipedia.org/wiki/Military%E2%80%93industrial_complex

[26]. Science History Institute. (2024, January 3). *History and Future of Plastics | Science History Institute*. Retrieved February 16, 2024, from https://sciencehistory.org/education/classroom-activities/role-playing-games/case-of-plastics/history-and-future-of-plastics/

[27]. *How highly processed foods liberated 1950s housewives*. (2017, May 11). National Women's History Museum. Retrieved March 10, 2024, from https://www.womenshistory.org/articles/how-highly-processed-foods-liberated-1950s-housewives

[28]. National Geographic. (2018, May 22). *A brief history of how plastic has changed our world | National Geographic* [Video]. YouTube. Retrieved February 16, 2024, from https://www.youtube.com/watch?v=jQdBag_p6kE

[29]. Johnstone, P., & McLeish, C. (2020). World wars and the age of oil: Exploring directionality in deep energy transitions. *Energy Research & Social Science, 69*, 101732. https://doi.org/10.1016/j.erss.2020.101732

[30]. Carbon Tracker. (2023, May 5). *The future's not in plastics - Carbon Tracker initiative*. Carbon Tracker Initiative. Retrieved March 10, 2024, from https://carbontracker.org/reports/the-futures-not-in-plastics/

[31]. *The plastics pipeline: A surge of new production is on the way*. (n.d.). Yale E360. Retrieved March 10, 2024, from https://e360.yale.edu/features/the-plastics-pipeline-a-surge-of-new-production-is-on-the-way

[32]. *A Fracking-Driven industrial boom renews pollution concerns in Pittsburgh*. (n.d.). Yale E360. Retrieved March 10, 2024, from https://e360.yale.edu/features/a-fracking-driven-industrial-boom-renews-pollution-concerns-in-pittsburgh

[33]. *ExxonMobil plans investments of $20 billion to expand manufacturing in U.S. Gulf region*. (2017, March 6). Exxon Mobil Corporation. Retrieved March 10, 2024, from https://investor.exxonmobil.com/news-events/press-releases/detail/386/exxonmobil-plans-investments-of-20-billion-to-expand

[34]. *Top oil and gas companies by revenue 2023 |*

Statista. (2023, August 29). Statista. Retrieved March 10, 2024, from https://www.statista.com/statistics/272710/top-10-oil-and-gas-companies-worldwide-based-on-revenue/
[35]. Wikipedia contributors. (2024a, February 10). *Fossil fuels lobby*. Wikipedia. Retrieved March 10, 2024, from https://en.wikipedia.org/wiki/Fossil_fuels_lobby
[36]. Matthews, M. A. and Shell International Chemical Company Ltd. "The Earth's Carbon Cycle." *The New Scientist*, October 8, 1959, 644–46. Accessed March 10, 2024. https://www.documentcloud.org/documents/4447739-The-Earth-s-Carbon-Cycle-Shell.
[37]. Union of Concerned Scientists. "The Disinformation Playbook," October 10, 2017. Accessed March 10, 2024. https://www.ucsusa.org/resources/disinformation-playbook; Wikipedia. "Tobacco Industry Playbook," March 23, 2023. Accessed March 10, 2024. https://en.wikipedia.org/wiki/Tobacco_industry_playbook

[38]

. Perkins, T. (2023, June 7). Chemical industry used big tobacco's tactics to conceal evidence of PFAS risks. *The Guardian*. Retrieved March 10, 2024, from https://www.theguardian.com/environment/2023/jun/07/pfas-3m-dupont-chemical-industry-health-toxic-study
[39]. *Front page | Minnesota 3M PFAS Settlement*. (n.d.). Retrieved March 10, 2024, from https://3msettlement.state.mn.us/
[40]. *For decades, polluters knew PFAS chemicals were dangerous but hid risks from public (Research)*. (2019b, August 29). Environmental Working Group. Retrieved March 10, 2024, from https://www.ewg.org/research/decades-polluters-knew-pfas-chemicals-were-dangerous-hid-risks-public
[41]. *For decades, polluters knew PFAS chemicals were dangerous but hid risks from public*. (2019, August 28). Environmental Working Group. Retrieved March 10, 2024, from https://www.ewg.org/news-insights/news/decades-polluters-knew-pfas-chemicals-were-dangerous-hid-risks-public
[42]. Hayes, J. (2019). *State of Minnesota V. 3M Co.: Exhibit 1124*.

Environmental Working Group. Retrieved March 10, 2024, from https://static.ewg.org/reports/2019/pfa-timeline/1975_Dr-Guy.pdf?_gl=1*7z7zug*_gcl_au*ODIwNDA1Nzk0LjE3MTAxMjg3NTg.*_ga*Nzg2OTIzNDQ5LjE3MTAxMjg3NTg.*_ga_CS21GC49KT*MTcxMDEyODc1OC4xLjEuMTcxMDEzMjk0NC4wLjAuMA..&_ga=2.218501721.1334700989.1710128759-786923449.1710128758

[43]. Gaber, N., Bero, L., & Woodruff, T. J. (2023). The Devil they Knew: Chemical Documents Analysis of Industry Influence on PFAS Science. *Annals of Global Health, 89*(1). https://doi.org/10.5334/aogh.4013

[44]. Gladstone, S., & Gladstone, S. (2023, December 8). *New report: PFAS industry spent more than $110 million on lobbying since 2019*. Food & Water Watch. Retrieved March 10, 2024, from https://www.foodandwaterwatch.org/2023/11/07/new-report-pfas-industry-spent-more-than-110-million-on-lobbying-since-2019/; Perkins, T. (2023b, November 7). US chemical industry likely spent $110m trying to thwart PFAS legislation, study finds. *The Guardian*. Retrieved March 10, 2024, from https://www.theguardian.com/us-news/2023/nov/07/us-chemical-industry-110-million-thwart-pfas-legislation

[45]. Qian, N., Gao, X., Lang, X., Deng, H., Bratu, T. M., Chen, Q., Stapleton, P. A., Yan, B., & Min, W. (2024). Rapid single-particle chemical imaging of nanoplastics by SRS microscopy. *Proceedings of the National Academy of Sciences of the United States of America, 121*(3). https://doi.org/10.1073/pnas.2300582121

[46]. Elbein, S. (2024, January 9). The Hill. *The Hill*. Retrieved March 10, 2024, from https://thehill.com/policy/equilibrium-sustainability/4397670-bottled-water-industry-pushes-back-study-warning-nanoplastics/

[47]. Gelles, D., & Steel, E. (2021, October 20). How Chemours and DuPont avoid paying for PFAS pollution. *The New York Times*. Retrieved March 8, 2024, from https://www.nytimes.com/2021/10/20/business/chemours-dupont-pfas-genx-chemicals.html; The Harvard Law School Forum on Corporate Governance. (2020, November 27). *Environmental*

spinoffs: the attempt to dump liability through spin and bankruptcy. Retrieved March 8, 2024, from https://corpgov.law.harvard.edu/2020/11/27/environmental-spinoffs-the-attempt-to-dump-liability-through-spin-and-bankruptcy/; ICIS Explore. (2024, February 8). *Individual news - ICIS Explore.* Retrieved March 8, 2024, from https://www.icis.com/explore/resources/news/2020/10/15/10563956/dupont-used-chemours-dow-merger-to-shield-itself-from-pfas-claims-suit/

[48]. Perkins, T. (2024, January 30). A North Carolina PFAS factory claims its emissions fell by 99.99%. A Guardian test reveals otherwise. *The Guardian.* Retrieved March 10, 2024, from https://www.theguardian.com/us-news/2024/jan/28/north-carolina-pfas-forever-chemicals-testing

[49]. Wikipedia. "Information Council for the Environment," January 4, 2022. Accessed March 25, 2023. https://en.wikipedia.org/wiki/Information_Council_for_the_Environment.

[50]. Times, N. Y. (1988, August 30). Plastics industry, under pressure, begins to invest in recycling. *The New York Times.* Retrieved March 11, 2024, from https://www.nytimes.com/1988/08/30/science/plastics-industry-under-pressure-begins-to-invest-in-recycling.html

[51]. *The fraud of plastic recycling.* (n.d.). Retrieved March 11, 2024, from https://climateintegrity.org/plastics-fraud

[52]. FRONTLINE, PBS. (2021, April 15). Plastic Wars. *FRONTLINE.* Retrieved March 11, 2024, from https://www.pbs.org/wgbh/frontline/documentary/plastic-wars/

[53]. Sullivan, L. (2020, September 11). How big oil misled the public into believing plastic would be recycled. *NPR.* Retrieved March 11, 2024, from https://www.npr.org/2020/09/11/897692090/how-big-oil-misled-the-public-into-believing-plastic-would-be-recycled

[54]. Noor, D. (2024, February 16). 'They lied': plastics producers deceived public about recycling, report reveals. *The Guardian.* Retrieved March 11, 2024, from https://www.theguardian.com/us-news/2024/feb/15/recycling-plastics-producers-report

[55]. Elbein, S. (2023, November 22). The Hill. *The Hill.* Retrieved March 11, 2024, from https://thehill.com/policy/equilibrium-sustainability/4322021-fossil-fuel-industry-recycling-plastics/

[56]. *Toys Made with Recycled Plastics Expose Children to High Levels of Toxic Chemicals | IPEN.* (n.d.). Retrieved March 11, 2024, from https://ipen.org/news/toys-made-recycled-plastics-expose-children-high-levels-toxic-chemicals; Gayle, D. (2023, May 24). Recycled plastic can be more toxic and is no fix for pollution, Greenpeace warns. *The Guardian.* Retrieved March 11, 2024, from https://www.theguardian.com/environment/2023/may/24/recycled-plastic-more-toxic-no-fix-pollution-greenpeace-warns; Verma, R., Vinoda, K., Papireddy, M., & Gowda, A. (2016). Toxic Pollutants from Plastic Waste- A Review. *Procedia Environmental Sciences, 35*, 701–708. https://doi.org/10.1016/j.proenv.2016.07.069; Lloyd, C. (2023, May 26). Toxins hidden in plastics are the industry's dirty secret – recycling is not the answer. *The Guardian.* Retrieved March 11, 2024, from https://www.theguardian.com/commentisfree/2023/may/25/toxins-in-plastics-industrys-dirty-secret-recycling-not-answer

[57]. Wikipedia contributors. (2024, February 19). *Plastic.* Wikipedia. Retrieved February 20, 2024, from https://en.wikipedia.org/wiki/Plastic

[58]. Wikipedia contributors. (2023, August 24). *Mesoamerican rubber balls.* Wikipedia. Retrieved February 20, 2024, from https://en.wikipedia.org/wiki/Mesoamerican_rubber_balls

[59]. Wikipedia contributors. (2023a, July 8). *Vulcanization.* Wikipedia. Retrieved February 20, 2024, from https://en.wikipedia.org/wiki/Vulcanization

[60]. Wikipedia contributors. (2024a, January 11). *List of synthetic polymers.* Wikipedia. Retrieved February 20, 2024, from https://en.wikipedia.org/wiki/List_of_synthetic_polymers

[61]. Staff, H. (1942, August 1). *Plastics come of age.* Harper's Magazine. Retrieved February 20, 2024, from https://harpers.org/archive/1942/08/plastics-come-of-age/

[62]. Brigham, K. (2022, February 1). How the fossil fuel industry is pushing plastics on the world. *CNBC.* Retrieved February 20,

2024, from https://www.cnbc.com/2022/01/29/how-the-fossil-fuel-industry-is-pushing-plastics-on-the-world-.html
[63]. The Editors of Encyclopaedia Britannica. (2023, December 25). *Polymer | Description, Examples, Types, Material, Uses, & Facts*. Encyclopedia Britannica. Retrieved February 20, 2024, from https://www.britannica.com/science/polymer
[64]. The Editors of Encyclopaedia Britannica. (1998, July 20). *Monomer | Definition & Facts*. Encyclopedia Britannica. Retrieved February 20, 2024, from https://www.britannica.com/science/monomer
[65]. Wikipedia contributors. (2024a, January 20). *Cracking (chemistry)*. Wikipedia. Retrieved February 20, 2024, from https://en.wikipedia.org/wiki/Cracking_(chemistry)
[66]. Wikipedia contributors. (2015, November 24). *File:Plastic injection molding.webm - Wikipedia*. Retrieved February 20, 2024, from https://en.wikipedia.org/wiki/File:Plastic_Injection_Molding.webm
[67]. Grabianowski, E., & McManus, M. R. (2023, October 18). *What is Moonshine and how is Moonshine made?* HowStuffWorks. Retrieved February 20, 2024, from https://science.howstuffworks.com/innovation/edible-innovations/moonshine.htm
[68]. *Chlorine: exposure, decontamination, treatment | Chemical emergencies | CDC*. (n.d.). Retrieved February 20, 2024, from https://www.cdc.gov/chemicalemergencies/factsheets/chlorine.html#anchor_55011
[69]. Wikipedia contributors. (2024d, February 8). *Catalysis*. Wikipedia. Retrieved February 20, 2024, from https://en.wikipedia.org/wiki/Catalysis

[70]

. Wikipedia contributors. (2024c, February 8). *Polymerization*. Wikipedia. Retrieved February 20, 2024, from https://en.wikipedia.org/wiki/Polymerization
[71]. *Ethane cracker plants: What are they?* (2018, October 23). The Climate Reality Project. Retrieved February 20,

2024, from https://www.climaterealityproject.org/blog/ethane-cracker-plants-what-are-they
[72]. Wikipedia contributors. (2024b, January 11). *Resin identification code*. Wikipedia. Retrieved February 20, 2024, from https://en.wikipedia.org/wiki/Resin_identification_code
[73]. Xometry, T. (2022, December 31). Polypropylene vs. HDPE: Material Differences and Comparisons. *Xometry.* Retrieved March 4, 2024, from https://www.xometry.com/resources/materials/polypropylene-vs-hdpe/
[74]. Xometry, T. (2022a, December 31). Polycarbonate vs. PVC: Material Differences and Comparisons. *Xometry*. Retrieved March 4, 2024, from https://www.xometry.com/resources/materials/polycarbonate-vs-pvc/
[75]. *What are volatile organic compounds (VOCs)? | US EPA*. (2023, March 15). US EPA. Retrieved February 22, 2024, from https://www.epa.gov/indoor-air-quality-iaq/what-are-volatile-organic-compounds-vocs
[76]. Gray, A. (2020, February 12). *What you need to know about Off-Gassing. Architectural Digest.* Retrieved February 22, 2024, from https://www.architecturaldigest.com/story/what-is-off-gassing
[77]. Wikipedia contributors. (2024, February 22). *Melamine*. Wikipedia. Retrieved February 22, 2024, from https://en.wikipedia.org/wiki/Melamine
[78]. Wikipedia contributors. (2024a, February 19). *Acrylic*. Wikipedia. Retrieved February 22, 2024, from https://en.wikipedia.org/wiki/Acrylic
[79]. Wikipedia contributors. (2024, January 30). *Tritan copolyester*. Wikipedia. Retrieved February 21, 2024, from https://en.wikipedia.org/wiki/Tritan_copolyester
[80]. Wikipedia contributors. (2024c, February 17). *Silicone*. Wikipedia. Retrieved February 21, 2024, from https://en.wikipedia.org/wiki/Silicone
[81]. *Silicone*. (n.d.). Life Without Plastic. Retrieved February 21, 2024, from https://lifewithoutplastic.com/silicone/
[82]. Wikipedia contributors. (2024a, January 28). *Polytetrafluoroethylene*. Wikipedia. Retrieved February 21, 2024,

from https://en.wikipedia.org/wiki/Polytetrafluoroethylene
[83]. Fenton, S. E., Ducatman, A., Boobis, A. R., DeWitt, J. C., Lau, C., Ng, C. A., Smith, J. S., & Roberts, S. M. (2020). Per- and Polyfluoroalkyl Substance Toxicity and Human Health Review: Current State of Knowledge and Strategies for Informing Future Research. *Environmental Toxicology and Chemistry, 40*(3), 606–630. https://doi.org/10.1002/etc.4890
[84]. Wikipedia contributors. (2023, December 19). *Superhydrophobic coating*. Wikipedia. Retrieved February 21, 2024, from https://en.wikipedia.org/wiki/Superhydrophobic_coating

[85]

. Bell-Young, L. (2022, July 29). *The science behind a soda can*. ReAgent Chemical Services. Retrieved February 21, 2024, from https://www.reagent.co.uk/blog/the-science-behind-a-soda-can/
[86]. Dockrill, P. (2019, May 22). *Viral video reveals the bizarre way you can make a soda can fully transparent : ScienceAlert*. ScienceAlert. Retrieved February 21, 2024, from https://www.sciencealert.com/viral-video-reveals-the-bizarre-way-you-can-make-a-soda-can-fully-transparent
[87]. Wikipedia contributors. (2023, June 14). *Metallised film*. Wikipedia. Retrieved February 22, 2024, from https://en.wikipedia.org/wiki/Metallised_film
[88]. Van De Walle Ms Rd, G. (2020, June 4). *Can you microwave plastic?* Healthline. Retrieved February 22, 2024, from https://www.healthline.com/nutrition/can-you-microwave-plastic#safety
[89]. Wikipedia contributors. (2024a, January 29). *Microbead*. Wikipedia. Retrieved February 22, 2024, from https://en.wikipedia.org/wiki/Microbead
[90]. Kleida, D. (2020, July 8). *72% of Sun Care products contain microplastics*. Beat the Microbead. Retrieved February 22, 2024, from https://www.beatthemicrobead.org/72-of-sun-care-products-contain-microplastics/
[91]. Wikipedia contributors. (2024a, January 15). *Glitter*.

Wikipedia. Retrieved February 22, 2024, from https://en.wikipedia.org/wiki/Glitter

[92]. Van Uffelen, C. (2020, March 2). *Microplastics found in 119 detergent brands - Plastic Soup Foundation.* Plastic Soup Foundation. Retrieved March 11, 2024, from https://www.plasticsoupfoundation.org/en/2019/07/microplastics-found-in-119-detergent-brands/

[93]. United Nations Environment Programme. (n.d.). *Fashion's tiny hidden secret.* UNEP. Retrieved February 23, 2024, from https://www.unep.org/news-and-stories/story/fashions-tiny-hidden-secret

[94]. Reporter, G. S. (2023, July 3). Are your clothes making you sick? The opaque world of chemicals in fashion. *The Guardian.* Retrieved February 23, 2024, from https://www.theguardian.com/fashion/2023/jul/02/fashion-chemicals-pfas-bpa-toxic

[95]. McNeely, E., Staffa, S. J., Mordukhovich, I., & Coull, B. A. (2017). Symptoms related to new flight attendant uniforms. *BMC Public Health, 17*(1). https://doi.org/10.1186/s12889-017-4982-4

[96]. Wicker, A. (2023). *To dye for: How Toxic Fashion Is Making Us Sick--and How We Can Fight Back.* Penguin.

[97]. Wikipedia contributors. (2024, February 19). *Microplastics.* Wikipedia. Retrieved February 23, 2024, from https://en.wikipedia.org/wiki/Microplastics

[98]. Milne, M. H., De Frond, H., Rochman, C. M., Mallos, N. J., Leonard, G. H., & Baechler, B. R. (2024b). Exposure of U.S. adults to microplastics from commonly-consumed proteins. *Environmental Pollution, 343*, 123233. https://doi.org/10.1016/j.envpol.2023.123233; Elbein, S. (2024, January 30). The Hill. *The Hill.* Retrieved March 11, 2024, from https://thehill.com/policy/equilibrium-sustainability/4397696-microplastics-study-american-proteins-meat-fish-plants/

[99]. Liu, J., Liu, Q., An, L., Wang, M., Yang, Q., Zhu, B., Ding, J., Ye, C., & Xu, Y. (2022). Microfiber pollution in the Earth system. *Reviews of Environmental Contamination and Toxicology, 260*(1). https://doi.org/10.1007/s44169-022-00015-9

[100]. Napper, I. E., & Thompson, R. C. (2016). Release of synthetic microplastic plastic fibres from domestic washing machines: Effects of fabric type and washing conditions. *Marine Pollution Bulletin, 112*(1–2), 39–45. https://doi.org/10.1016/j.marpolbul.2016.09.025
[101]. Allen, E., Henninger, C. E., Garforth, A., & Asuquo, E. D. (2024). Microfiber Pollution: A Systematic Literature Review to overcome the complexities in knit design to create solutions for knit fabrics. *Environmental Science & Technology*. https://doi.org/10.1021/acs.est.3c05955
[102]. Perkins, T. (2023, August 17). 90% of Great Lakes water samples have unsafe microplastic levels – report. *The Guardian*. Retrieved March 11, 2024, from https://www.theguardian.com/environment/2023/aug/17/great-lakes-water-microplastic-pollution-contamination
[103]. Ding, J., Sun, C., He, C., Zheng, L., Dai, D., & Li, F. (2022). Atmospheric microplastics in the Northwestern Pacific Ocean: Distribution, source, and deposition. *Science of the Total Environment, 829*, 154337. https://doi.org/10.1016/j.scitotenv.2022.154337; Allen, S., Allen, D., Baladima, F., Phoenix, V. R., Thomas, J. L., Roux, G. L., & Sonke, J. E. (2021). Evidence of free tropospheric and long-range transport of microplastic at Pic du Midi Observatory. *Nature Communications, 12*(1). https://doi.org/10.1038/s41467-021-27454-7
[104]. Wang, Y., Okochi, H., Tani, Y., Hayami, H., Minami, Y., Katsumi, N., Takeuchi, M., Sorimachi, A., Fujii, Y., Kajino, M., Adachi, K., Ishihara, Y., Iwamoto, Y., & Niida, Y. (2023). Airborne hydrophilic microplastics in cloud water at high altitudes and their role in cloud formation. *Environmental Chemistry Letters, 21*(6), 3055–3062. https://doi.org/10.1007/s10311-023-01626-x; Perkins, T. (2023, October 9). Microplastics detected in clouds hanging atop two Japanese mountains. *The Guardian*. Retrieved February 23, 2024, from https://www.theguardian.com/environment/2023/oct/09/microplastics-clouds-study-mount-fuji-mount-oyama
[105]. Fan, W., Salmond, J., Dirks, K. N., Sanz, P. C., Miskelly, G.

M., & Rindelaub, J. D. (2022). Evidence and mass quantification of atmospheric microplastics in a coastal New Zealand city. *Environmental Science & Technology, 56*(24), 17556–17568. https://doi.org/10.1021/acs.est.2c05850

[106]. Amato-Lourenço, L. F., Carvalho-Oliveira, R., Júnior, G. R., Galvão, L. D. S., Ando, R. A., & Mauad, T. (2021). Presence of airborne microplastics in human lung tissue. *Journal of Hazardous Materials, 416*, 126124. https://doi.org/10.1016/j.jhazmat.2021.126124; Carrington, D. (2022, April 7). Microplastics found deep in lungs of living people for first time. *The Guardian*. Retrieved February 23, 2024, from https://www.theguardian.com/environment/2022/apr/06/microplastics-found-deep-in-lungs-of-living-people-for-first-time

[107]. Leslie, H., Van Velzen, M., Brandsma, S., Vethaak, D., García-Vallejo, J. J., & Lamoree, M. (2022). Discovery and quantification of plastic particle pollution in human blood. *Environment International, 163*, 107199. https://doi.org/10.1016/j.envint.2022.107199; Carrington, D. (2022a, March 25). Microplastics found in human blood for first time. *The Guardian*. Retrieved February 23, 2024, from https://www.theguardian.com/environment/2022/mar/24/microplastics-found-in-human-blood-for-first-time

[108]. Jenner, L. C., Rotchell, J. M., Bennett, R. T., Cowen, M. E., Tentzeris, V., & Sadofsky, L. R. (2022). Detection of microplastics in human lung tissue using μFTIR spectroscopy. *Science of the Total Environment, 831*, 154907. https://doi.org/10.1016/j.scitotenv.2022.154907

[109]. Simon, M. (2020b, June 11). Plastic rain is the new acid rain. *WIRED*. Retrieved February 25, 2024, from https://www.wired.com/story/plastic-rain-is-the-new-acid-rain/; *Scientists find plastic pollution in the rain and in the air we breathe*. (2020, October 5). World Economic Forum. Retrieved February 25, 2024, from https://www.weforum.org/agenda/2020/07/scientists-find-plastic-pollution-in-the-rain-and-in-the-air-we-breathe/; Brahney, J., Hallerud, M., Heim, E., Hahnenberger, M., &

Sukumaran, S. (2020). Plastic rain in protected areas of the United States. *Science, 368*(6496), 1257–1260. https://doi.org/10.1126/science.aaz5819

[110]. Gao, Z., Wontor, K., Cizdziel, J. V., & Lü, H. (2022). Distribution and characteristics of microplastics in beach sand near the outlet of a major reservoir in north Mississippi, USA. *Microplastics and Nanoplastics, 2*(1). https://doi.org/10.1186/s43591-022-00029-z

[111]. *Ocean trash: 5.25 trillion pieces and counting, but big questions remain.* (n.d.). Retrieved February 25, 2024, from https://education.nationalgeographic.org/resource/ocean-trash-525-trillion-pieces-and-counting-big-questions-remain/

[112]. Race, O. (n.d.). *The Ocean Race 2022-23.* The Ocean Race 2022-23. Retrieved February 25, 2024, from https://www.theoceanrace.com/

[113]. Reporter, G. S. (2023, June 6). Microplastics found in every sample of water taken during the Ocean Race. *The Guardian.* Retrieved February 25, 2024, from https://www.theguardian.com/environment/2023/jun/05/microplastics-found-in-every-sample-of-water-taken-during-ocean-race; Reporter, G. S. (2023a, March 28). 'It's win-win': how a dangerous sailing race could reveal the ocean's secrets. *The Guardian.* Retrieved February 25, 2024, from https://www.theguardian.com/environment/2023/mar/25/southern-ocean-race-science-sailing

[114]. Carrington, D. (2018, October 22). Plastic fibres found in tap water around the world, study reveals. *The Guardian.* Retrieved February 25, 2024, from https://www.theguardian.com/environment/2017/sep/06/plastic-fibres-found-tap-water-around-world-study-reveals

[115]. Common, D., & Szeto, E. (2018, April 6). Microplastics found in 93% of bottled water tested in global study. *CBC.* Retrieved February 25, 2024, from https://www.cbc.ca/news/science/bottled-water-microplastics-1.4575045

[116]. Doubek, J. (2024b, January 10). Researchers find a massive number of plastic particles in bottled water. *NPR.* Retrieved February 25, 2024,

from https://www.npr.org/2024/01/10/1223730333/bottled-water-plastic-microplastic-nanoplastic-study; Qian, N., Gao, X., Lang, X., Deng, H., Bratu, T. M., Chen, Q., Stapleton, P. A., Yan, B., & Min, W. (2024). Rapid single-particle chemical imaging of nanoplastics by SRS microscopy. *Proceedings of the National Academy of Sciences of the United States of America, 121*(3). https://doi.org/10.1073/pnas.2300582121

[117]. Li, D., Shi, Y., Yang, L., Xiao, L., Kehoe, D. K., Gun'ko, Y. K., Boland, J. J., & Wang, J. J. (2020). Microplastic release from the degradation of polypropylene feeding bottles during infant formula preparation. *Nature Food, 1*(11), 746–754. https://doi.org/10.1038/s43016-020-00171-y; Carrington, D. (2020, October 20). Bottle-fed babies swallow millions of microplastics a day, study finds. The *Guardian*. Retrieved February 25, 2024, from https://www.theguardian.com/environment/2020/oct/19/bottle-fed-babies-swallow-millions-microplastics-day-study

[118]. Ragusa, A., Svelato, A., Santacroce, C., Catalano, P., Notarstefano, V., Carnevali, O., Papa, F., Rongioletti, M., Baiocco, F., Draghi, S., D'Amore, E., Rinaldo, D., Matta, M., & Giorgini, E. (2021b). Plasticenta: First evidence of microplastics in human placenta. *Environment International, 146*, 106274. https://doi.org/10.1016/j.envint.2020.106274; Carrington, D. (2020b, December 23). Microplastics revealed in the placentas of unborn babies. *The Guardian*. Retrieved February 25, 2024, from https://www.theguardian.com/environment/2020/dec/22/microplastics-revealed-in-placentas-unborn-babies

[119]. Ragusa, A., Notarstefano, V., Svelato, A., Belloni, A., Gioacchini, G., Blondeel, C., Zucchelli, E., De Luca, C., D'Avino, S., Gulotta, A., Carnevali, O., & Giorgini, E. (2022). Raman Microspectroscopy Detection and characterisation of microplastics in human breastmilk. *Polymers, 14*(13), 2700. https://doi.org/10.3390/polym14132700; Carrington, D. (2022, October 7). Microplastics found in human breast milk for the first time. *The Guardian*. Retrieved February 25, 2024, from https://www.theguardian.com/environment/2022/oct/07/microplastics-human-breast-milk-first-time

[120]. Jadhav, E. B., Sankhla, M. S., Bhat, R. A., & Bhagat, D. S. (2021). Microplastics from food packaging: An overview of human consumption, health threats, and alternative solutions. *Environmental Nanotechnology, Monitoring and Management, 16*, 100608. https://doi.org/10.1016/j.enmm.2021.100608

[121]. Images, U. I. G. (2018, October 17). A new study looked at sea, rock, and lake salt sold around the world. Here's what you need to know. *National Geographic*. Retrieved February 25, 2024, from https://www.nationalgeographic.co.uk/environment-and-conservation/2018/10/microplastics-found-in-90-percent-of-table-salt; Kim, J., Lee, H., Kim, S., & Kim, H. (2018). Global Pattern of Microplastics (MPS) in Commercial Food-Grade Salts: Sea salt as an indicator of Seawater MP Pollution. *Environmental Science & Technology, 52*(21), 12819–12828. https://doi.org/10.1021/acs.est.8b04180

[122]. Diaz-Basantes, M. F., Nacimba-Aguirre, D., Conesa, J. A., & Fullana, A. (2022). Presence of microplastics in commercial canned tuna. *Food Chemistry, 385*, 132721. https://doi.org/10.1016/j.foodchem.2022.132721

[123]. *Microplastics in food commodities*. (2022). In FAO eBooks. https://doi.org/10.4060/cc2392en

[124]. Garw. (2019, June 13). *How much microplastics are we ingesting?: Estimation of the mass of microplastics ingested*. The University of Newcastle, Australia. Retrieved February 25, 2024, from https://www.newcastle.edu.au/newsroom/featured/plastic-ingestion-by-people-could-be-equating-to-a-credit-card-a-week/how-much-microplastics-are-we-ingesting-estimation-of-the-mass-of-microplastics-ingested

[125]. Scarr, S. (n.d.). A plateful of plastic. *Reuters*. Retrieved February 25, 2024, from https://www.reuters.com/graphics/ENVIRONMENT-PLASTIC/0100B4TF2MQ/index.html

[126]. Lai, H., Liu, X., & Qu, M. (2022). Nanoplastics and human Health: hazard identification and biointerface. *Nanomaterials, 12*(8), 1298. https://doi.org/10.3390/nano12081298

[127]. The "Plasticene" epoch? (2018). *Elements, 14*(5), 291. https://doi.org/10.2138/gselements.14.5.291

[128]. *Global plastics production.* (n.d.-b). Our World in Data. Retrieved March 13, 2024, from https://ourworldindata.org/grapher/global-plastics-production
[129]. *Cumulative global production of plastics.* (n.d.-b). Our World in Data. Retrieved March 13, 2024, from https://ourworldindata.org/grapher/cumulative-global-plastics
[130]. Elhacham, E., Ben-Uri, L., Grozovski, J., & Bar-On, Y. M. (2020). Global human-made mass exceeds all living biomass. *Nature, 588*(7838), 442–444. https://doi.org/10.1038/s41586-020-3010-5
[131]. *Plastic pollution.* (n.d.-b). Our World in Data. Retrieved March 13, 2024, from https://slides.ourworldindata.org/plastic-pollution/#/5
[132]. *Annual global plastic waste generation by industrial sector.* (n.d.). Our World in Data. Retrieved March 13, 2024, from https://ourworldindata.org/grapher/plastic-waste-by-sector
[133]. *Global primary plastic waste generation by polymer.* (n.d.). Our World in Data. Retrieved March 13, 2024, from https://ourworldindata.org/grapher/plastic-waste-polymer
[134]. Brooks, A., Wang, S., & Jambeck, J. (2018). The Chinese import ban and its impact on global plastic waste trade. *Science Advances, 4*(6). https://doi.org/10.1126/sciadv.aat0131
[135]. Whitehouse, D. (2020, November 19). Kenya is not a dumping ground for US plastic. *The Africa Report.com.* Retrieved March 13, 2024, from https://www.theafricareport.com/51072/kenya-is-not-a-dumping-ground-for-us-plastic/

[136]

. Calderwood, I. (2018, April 25). 16 *times countries and cities have banned Single-Use plastics*. Global Citizen. Retrieved March 13, 2024, from https://www.globalcitizen.org/en/content/plastic-bans-around-the-world/
[137]. Mpungu, P. (2020, December 4). Despite pioneering ban, Kenya is drowning in single-use plastic. *Al Jazeera.* Retrieved March 13, 2024, from https://www.aljazeera.com/economy/2020/12/4/despite-pioneering-ban-kenya-is-

drowning-in-single-use-plastic
[138]. Wikipedia contributors. (2024, March 12). *Basel Convention*. Wikipedia. Retrieved March 13, 2024, from https://en.wikipedia.org/wiki/Basel_Convention
[139]. Basel Convention. (n.d.). *Parties to the Basel Convention.* Copyright 2011 by Basel Convention. Retrieved March 13, 2024, from https://www.basel.int/Countries/StatusofRatifications/PartiesSignatories/tabid/4499/Default.aspx
[140]. Tabuchi, H., & Corkery, M. (2021, March 12). Countries tried to curb trade in plastic waste. the U.S. is shipping more. *The New York Times.* Retrieved March 13, 2024, from https://www.nytimes.com/2021/03/12/climate/plastics-waste-export-ban.html
[141]. *Plastics in the Ocean - Ocean Conservancy.* (n.d.). Ocean Conservancy. Retrieved March 13, 2024, from https://oceanconservancy.org/trash-free-seas/plastics-in-the-ocean/
[142]. Rannard, B. G. (2023, March 8). Oceans littered with 171 trillion plastic pieces. *BBC*. Retrieved March 13, 2024, from https://www.bbc.com/news/science-environment-64889284
[143]. Mishra, S., Rath, C. C., & Das, A. P. (2019). Marine microfiber pollution: A review on present status and future challenges. *Marine Pollution Bulletin, 140*, 188–197. https://doi.org/10.1016/j.marpolbul.2019.01.039
[144]. Wikipedia contributors. (2024a, February 19). *Ocean gyre*. Wikipedia. Retrieved March 13, 2024, from https://en.wikipedia.org/wiki/Ocean_gyre
[145]. *Great Pacific Garbage patch.* (n.d.). National Geographic. Retrieved March 13, 2024, from https://education.nationalgeographic.org/resource/great-pacific-garbage-patch/
[146]. Wikipedia contributors. (2024b, March 5). *Great Pacific garbage patch*. Wikipedia. Retrieved March 13, 2024, from https://en.wikipedia.org/wiki/Great_Pacific_garbage_patch
[147]. Quecholac-Piña, X., Del Consuelo Hernández-Berriel, M., Del Consuelo Mañón-Salas, M., Espinosa-Valdemar, R. M., & Vázquez Morillas, A. (2020). Degradation of Plastics under Anaerobic

Conditions: A Short Review. *Polymers, 12*(1), 109. https://doi.org/10.3390/polym12010109
[148]. RTS - Recycle Track Systems. (2024, February 20). *Is recycling worth it? Costs and benefits of recycling | RTS*. Recycle Track Systems. Retrieved March 14, 2024, from https://www.rts.com/blog/is-recycling-worth-it/

[149]

. Ambrose, J. (2020, September 23). War on plastic waste faces setback as cost of recycled material soars. *The Guardian*. Retrieved March 14, 2024, from https://www.theguardian.com/environment/2019/oct/13/war-on-plastic-waste-faces-setback-as-cost-of-recycled-material-soars
[150]. Aktürk, N., & Aktürk, N. (2024, February 21). *28 Statistics about clothing donation & the used / second hand resale market*. Threads Monthly |. Retrieved March 14, 2024, from https://threadsmonthly.com/clothing-donation-stats/#dont-sell; Schumacher, K. A. (2022). *Facilitating a Circular Economy for Textiles workshop report*. https://doi.org/10.6028/nist.sp.1500-207
[151]. McVeigh, K. (2022, October 19). Huge rise in US plastic waste shipments to poor countries following China ban. *The Guardian*. Retrieved March 14, 2024, from https://www.theguardian.com/global-development/2018/oct/05/huge-rise-us-plastic-waste-shipments-to-poor-countries-china-ban-thailand-malaysia-vietnam
[152]. Sullivan, L. (2022, October 24). Recycling plastic is practically impossible — and the problem is getting worse. *NPR*. Retrieved March 14, 2024, from https://www.npr.org/2022/10/24/1131131088/recycling-plastic-is-practically-impossible-and-the-problem-is-getting-worse; Gregory, T. (2022, October 26). *Circular claims fall flat again - Greenpeace USA*. Greenpeace USA - We Fight for a Greener, More Peaceful World. Retrieved March 14, 2024, from https://www.greenpeace.org/usa/reports/circular-claims-

fall-flat-again/

[153]. Perkins, T. (2023, May 28). Recycled and reused food contact plastics are 'vectors' for toxins – study. *The Guardian.* Retrieved March 14, 2024, from https://www.theguardian.com/environment/2023/may/27/recycled-reused-food-plastic-toxins-study

[154]. Geueke, B., Phelps, D., Parkinson, L. V., & Muncke, J. (2023). Hazardous chemicals in recycled and reusable plastic food packaging. *Cambridge Prisms Plastics*, 1–43. https://doi.org/10.1017/plc.2023.7

[155]. Brooks, T. (2024, March 7). *New Greenpeace report calls out toxic hazards of recycled plastic as Global Plastics Treaty negotiations resume in Paris - Greenpeace USA*. Greenpeace USA - We Fight for a Greener, More Peaceful World. Retrieved March 14, 2024, from https://www.greenpeace.org/usa/news/new-greenpeace-report-calls-out-toxic-hazards-of-recycled-plastic-as-global-plastics-treaty-negotiations-resume-in-paris/

[156]. Wikipedia contributors. (2024, February 26). *Plastic recycling.* Wikipedia. Retrieved March 14, 2024, from https://en.wikipedia.org/wiki/Plastic_recycling#Energy_recovery

[157]. Staff, Z. (2021, August 30). *The story of REC: The hidden emissions of the youngest Dutch incinerator - Zero Waste Europe.* Zero Waste Europe. Retrieved March 14, 2024, from https://zerowasteeurope.eu/2019/06/the-story-of-rec/; *The True Toxic Toll: Biomonitoring of incineration emissions - Zero Waste Europe.* (2022, September 15). Zero Waste Europe. Retrieved March 14, 2024, from https://zerowasteeurope.eu/library/the-true-toxic-toll-biomonitoring-of-incineration-emissions/

[158]. Wikipedia contributors. (2024, February 25). *Polylactic acid.* Wikipedia. Retrieved March 12, 2024, from https://en.wikipedia.org/wiki/Polylactic_acid

[159]

. Harris, W. (2023, September 8). *How long does it take for plastic to decompose?* HowStuffWorks. Retrieved March 12, 2024, from https://science.howstuffworks.com/science-vs-myth/everyday-myths/how-long-does-it-take-for-plastics-to-biodegrade.htm

[160]. Rebecca Hale and Mark Thiessen, National Geographic. (2018, November 16). Can bioplastics truly relieve pressure on the environment? Experts weigh in. *National Geographic*. Retrieved March 12, 2024, from https://www.nationalgeographic.co.uk/environment-and-conservation/2018/11/everything-you-need-to-know-about-plant-based-plastics

[161]. *New process makes 'biodegradable' plastics truly compostable.* (2021, April 21). Berkeley. Retrieved March 16, 2024, from https://news.berkeley.edu/2021/04/21/new-process-makes-biodegradable-plastics-truly-compostable/

[162]. *The truth about bioplastics.* (2017, December 13). State of the Planet. Retrieved March 12, 2024, from https://news.climate.columbia.edu/2017/12/13/the-truth-about-bioplastics/

[163]. Horton, H. (2023, May 10). Microbes discovered that can digest plastics at low temperatures. *The Guardian*. Retrieved March 14, 2024, from https://www.theguardian.com/science/2023/may/10/microbes-digest-plastics-low-temperatures-recycling

[164]. Rüthi, J., Cerri, M., Brunner, I., Stierli, B., Sander, M., & Frey, B. (2023). Discovery of plastic-degrading microbial strains isolated from the alpine and Arctic terrestrial plastisphere. *Frontiers in Microbiology, 14*. https://doi.org/10.3389/fmicb.2023.1178474

[165]. Buranyi, S. (2023, November 3). 'We are just getting started': the plastic-eating bacteria that could change the world. *The Guardian*. Retrieved March 14, 2024, from https://www.theguardian.com/environment/2023/sep/28/plastic-eating-bacteria-enzyme-recycling-waste

[166]. Wikipedia contributors. (2024b, February 29). *Ideonella sakaiensis*. Wikipedia. Retrieved March 14, 2024, from https://en.wikipedia.org/wiki/Ideonella_sakaiensis

[167]. Yoshida, S., Hiraga, K., Takehana, T., Taniguchi, I., Yamaji, H., Maeda, Y., Toyohara, K., Miyamoto, K., Kimura, Y., & Oda, K. (2016). A bacterium that degrades and assimilates

poly(ethylene terephthalate). *Science, 351*(6278), 1196–1199. https://doi.org/10.1126/science.aad6359

[168]. Dutfield, S. (2022, March 23). *Plastic-eating bacteria: Genetic engineering and environmental impact.* livescience.com. Retrieved March 14, 2024, from https://www.livescience.com/plastic-eating-bacteria

[169]. Earth911.com. (2021, March 13). *How to recycle tin or steel cans.* Earth911. Retrieved March 18, 2024, from https://earth911.com/recycling-guide/how-to-recycle-tin-or-steel-cans/

[170]. Kannan, K., & Vimalkumar, K. (2021). A review of human exposure to microplastics and insights into microplastics as obesogens. *Frontiers in Endocrinology, 12.* https://doi.org/10.3389/fendo.2021.724989

[171]. Watson, C. (2024, March 7). *Plastic found inside more than 50% of plaques from clogged arteries : ScienceAlert.* ScienceAlert. Retrieved March 7, 2024, from https://www.sciencealert.com/plastic-found-inside-more-than-50-of-plaques-from-clogged-arteries; Kozlov, M., & Magazine, N. (2024, March 9). *Microplastics linked to heart attack, stroke and death.* Scientific American. Retrieved March 11, 2024, from https://www.scientificamerican.com/article/microplastics-linked-to-heart-attack-stroke-and-death/; Marfella, R., Prattichizzo, F., Sardu, C., Fulgenzi, G., et al. (2024). Microplastics and nanoplastics in atheromas and cardiovascular events. *The New England Journal of Medicine, 390*(10), 900–910. https://doi.org/10.1056/NEJMoa2309822

[172]. Prüst, M., Meijer, J., & Westerink, R. H. (2020). The plastic brain: neurotoxicity of micro- and nanoplastics. *Particle and Fibre Toxicology, 17*(1). https://doi.org/10.1186/s12989-020-00358-y

[173]. Collins, C. (2023, September 4). STARTLInG Findings – Scientists discover that microplastics could be changing your brain. *SciTechDaily.* Retrieved March 5, 2024, from https://scitechdaily.com/startling-findings-scientists-discover-that-microplastics-could-be-changing-your-brain/

[174]. Loria, K. (2019, September 17). Most plastic products contain potentially toxic chemicals, study reveals. *Consumer Reports.*

https://www.consumerreports.org/toxic-chemicals-substances/most-plastic-products-contain-potentially-toxic-chemicals/

[175]. Zimmermann, L., Dierkes, G., Ternes, T. A., Völker, C., & Wagner, M. (2019). Benchmarking the in vitro toxicity and chemical composition of plastic consumer products. *Environmental Science & Technology, 53*(19), 11467–11477. https://doi.org/10.1021/acs.est.9b02293

[176]. Wiesinger, H., Wang, Z., & Hellweg, S. (2021). Deep Dive into Plastic Monomers, Additives, and Processing Aids. *Environmental Science & Technology, 55*(13), 9339–9351. https://doi.org/10.1021/acs.est.1c00976

[177]. Wikipedia contributors. (2024, January 3). *CAS Registry Number*. Wikipedia. Retrieved March 5, 2024, from https://en.wikipedia.org/wiki/CAS_Registry_Number

[178]. *Chemicals in Plastics - a technical report.* (n.d.). UNEP - UN Environment Programme. Retrieved March 4, 2024, from https://www.unep.org/resources/report/chemicals-plastics-technical-report

[179]. Wikipedia contributors. (2024, March 4). *Philosopher's stone*. Wikipedia. Retrieved March 7, 2024, from https://en.wikipedia.org/wiki/Philosopher's_stone

[180]. Zanolli, L. (2019, July 2). Bisphenol: what to know about the chemicals in water bottles and cans. *The Guardian.* Retrieved March 5, 2024, from https://www.theguardian.com/us-news/2019/may/24/bisphenols-bpa-everyday-toxics-guide

[181]. *The facts about Bisphenol A.* (2010, August 9). WebMD. Retrieved March 6, 2024, from https://www.webmd.com/children/bpa

[182]

. Belvoir Media Group. (2021, February 19). *BPA: How Worried Should We Be? - Tufts Health & Nutrition Letter.* Tufts Health & Nutrition Letter. Retrieved March 6, 2024, from https://www.nutritionletter.tufts.edu/healthy-eating/bpa-how-worried-should-we-be/

[183]. *France releases guidance on BPA ban and Food-Contact*

legislation | PackagingLaw.com. (2014, August 22). Retrieved March 6, 2024, from https://www.packaginglaw.com/news/france-releases-guidance-bpa-ban-and-food-contact-legislation

[184]. European Food Safety Authority. (2023b, December 1). *Bisphenol A*. Retrieved March 6, 2024, from https://www.efsa.europa.eu/en/topics/topic/bisphenol

[185]. Service, R. F. (2018). BPA substitutes may be just as bad as the popular consumer plastic. *Science*. https://doi.org/10.1126/science.aav4159

[186]. *Learn about Polychlorinated Biphenyls | US EPA*. (2023, April 12). US EPA. Retrieved March 6, 2024, from https://www.epa.gov/pcbs/learn-about-polychlorinated-biphenyls

[187]. Wang, Y., & Qian, H. (2021). Phthalates and their impacts on human health. *Healthcare, 9*(5), 603. https://doi.org/10.3390/healthcare9050603

[188]. Wikipedia contributors. (2024, February 16). *Phthalates*. Wikipedia. Retrieved March 6, 2024, from https://en.wikipedia.org/wiki/Phthalates

[189]. *Phthalates - ECHA*. (n.d.). Retrieved March 6, 2024, from https://www.echa.europa.eu/web/guest/hot-topics/phthalates

[190]. Nutrition, C. F. F. S. a. A. (2023, July 21). *Phthalates in food packaging and food contact applications*. U.S. Food And Drug Administration. Retrieved March 6, 2024, from https://www.fda.gov/food/food-ingredients-packaging/phthalates-food-packaging-and-food-contact-applications

[191]. Friedman, L. F. (2024, January 4). The plastic chemicals hiding in your food. *Consumer Reports*. Retrieved March 6, 2024, from https://www.consumerreports.org/health/food-contaminants/the-plastic-chemicals-hiding-in-your-food-a7358224781/

[192]. Wikipedia contributors. (2024a, January 29). *Xenoestrogen*. Wikipedia. Retrieved March 6, 2024, from https://en.wikipedia.org/wiki/Xenoestrogen

[193]. Wikipedia contributors. (2023, December 17). *Nonylphenol*. Wikipedia. Retrieved March 6, 2024, from https://en.wikipedia.org/wiki/Nonylphenol

[194]. Grün, F., & Blumberg, B. (2009). Endocrine disrupters as

obesogens. *Molecular and Cellular Endocrinology, 304*(1–2), 19–29. https://doi.org/10.1016/j.mce.2009.02.018

[195]. Yen, C., Sun, C., Leu, S., Wallace, C. G., Lin, Y., Chang, L., Chen, Y., Tsa, T., Kao, Y., Shao, P., Hsieh, C., Chen, Y., & Yip, H. (2012). Continuing exposure to Low-Dose nonylphenol aggravates Adenine-Induced chronic renal dysfunction and role of rosuvastatin therapy. *Journal of Translational Medicine, 10*(1). https://doi.org/10.1186/1479-5876-10-147

[196]

. Soares, A., Guieysse, B., Jefferson, B., Cartmell, E., & Lester, J. (2008). Nonylphenol in the environment: A critical review on occurrence, fate, toxicity and treatment in wastewaters. *Environment International*, 34(7), 1033–1049. https://doi.org/10.1016/j.envint.2008.01.004

[197]. Bechi, N., Ietta, F., Romagnoli, R., Jantra, S., Cencini, M., Galassi, G., Serchi, T., Corsi, I., Focardi, S., & Paulesu, L. (2010). Environmental Levels of para -Nonylphenol Are Able to Affect Cytokine Secretion in Human Placenta. *Environmental Health Perspectives, 118*(3), 427–431. https://doi.org/10.1289/ehp.0900882

[198]. Jubendradass, R., D'Cruz, S., & Mathur, P. P. (2011). Long-term exposure to nonylphenol affects insulin signaling in the liver of adult male rats. *Human & Experimental Toxicology, 31*(9), 868–876. https://doi.org/10.1177/0960327111426587

[199]. *Restriction of Textile Products Containing Nonylphenol and Its Ethoxylates Begins to Apply EU-wide*. (n.d.). HKTDC Research. Retrieved March 6, 2024, from https://research.hktdc.com/en/article/NjczMjg5NzY1

[200]. *Fact Sheet: Nonylphenols and nonylphenol ethoxylates | US EPA*. (2023b, June 8). US EPA. Retrieved March 6, 2024, from https://www.epa.gov/assessing-and-managing-chemicals-under-tsca/fact-sheet-nonylphenols-and-nonylphenol-ethoxylates

[201]. Wikipedia contributors. (2024a, February 8). *Superabsorbent polymer*. Wikipedia. Retrieved March 8, 2024, from https://en.wikipedia.org/wiki/Superabsorbent_polymer

[202]. Wikipedia contributors. (2024b, March 7). *Per- and polyfluoroalkyl substances.* Wikipedia. Retrieved March 8, 2024, from https://en.wikipedia.org/wiki/Per-_and_polyfluoroalkyl_substances
[203]. *The history of Teflon® fluoropolymers.* (n.d.). Retrieved March 8, 2024, from https://www.teflon.com/en/news-events/history
[204]. Kandheri, N., & Kandheri, N. (2023, September 27). *What is PFOA and Why it is Banned in EU REACH | APA Engineering.* APA Engineering |. Retrieved March 8, 2024, from https://apaengineering.com/compliance-blog/eu-pfoa-ban/; Hogue, C. (2019, May 7). *Governments endorse global PFOA ban, with some exemptions.* Chemical & Engineering News. Retrieved March 8, 2024, from https://cen.acs.org/environment/persistent-pollutants/Governments-endorse-global-PFOA-ban/97/web/2019/05
[205]. Salvidge, R., & Hosea, L. (2023, February 23). What are PFAS, how toxic are they and how do you become exposed? *The Guardian.* Retrieved March 8, 2024, from https://www.theguardian.com/environment/2023/feb/23/what-are-pfas-forever-chemicals-how-toxic-are-they-and-how-do-you-become-exposed; Bradshaw, P. (2020, April 16). *Dark Waters review – Todd Haynes plumbs the depths of a poisoning scandal.* The Guardian. Retrieved March 8, 2024, from https://www.theguardian.com/film/2020/feb/27/dark-waters-review-todd-haynes-mark-ruffalo-west-virginia-dupont-poison-; Rich, N. (2016, January 6). The lawyer who became DuPont's worst nightmare. *The New York Times Magazine.* Retrieved March 8, 2024, from https://www.nytimes.com/2016/01/10/magazine/the-lawyer-who-became-duponts-worst-nightmare.html
[206]. Wikipedia contributors. (2024d, March 6). *Persistent organic pollutant.* Wikipedia. Retrieved March 8, 2024, from https://en.wikipedia.org/wiki/Persistent_organic_pollutant
[207]. *What are PFAS chemicals?* (n.d.). Environmental Working Group. Retrieved March 8, 2024, from https://www.ewg.org/what-are-pfas-chemicals
[208]. PFAS in clothing: Is what you wear dripping in "forever

chemicals"? (2023, March 30). *CBS News*. Retrieved March 8, 2024, from https://www.cbsnews.com/news/pfas-in-clothing-household-items-consumer-products-forever-chemicals/

[209]. Perkins, T. (2022, January 5). Toxic 'forever chemicals' are contaminating plastic food containers. *The Guardian*. Retrieved March 8, 2024, from https://www.theguardian.com/environment/2021/jul/09/toxic-forever-chemicals-plastic-food-containers

[210]. *PFAS exposure and risk of cancer.* (n.d.). National Cancer Institute. Retrieved March 8, 2024, from https://dceg.cancer.gov/research/what-we-study/pfas

[211]. Rickard, B. P., Rizvi, I., & Fenton, S. E. (2022). Per- and poly-fluoroalkyl substances (PFAS) and female reproductive outcomes: PFAS elimination, endocrine-mediated effects, and disease. *Toxicology, 465*, 153031. https://doi.org/10.1016/j.tox.2021.153031

[212]. *PFAS chemicals harm the immune system, decrease response to vaccines, new EWG review finds.* (2019, June 21). Environmental Working Group. Retrieved March 8, 2024, from https://www.ewg.org/news-insights/news/pfas-chemicals-harm-immune-system-decrease-response-vaccines-new-ewg-review

[213]. Coperchini, F., Croce, L., Ricci, G., Magri, F., Rotondi, M., Imbriani, M., & Chiovato, L. (2021). Thyroid Disrupting Effects of Old and New Generation PFAS. *Frontiers in Endocrinology, 11*. https://doi.org/10.3389/fendo.2020.612320

[214]. *Exposure to PFAS associated with increased blood lipids, possible CVD risk.* (2023, May 31). News. Retrieved March 8, 2024, from https://www.hsph.harvard.edu/news/hsph-in-the-news/exposure-to-pfas-associated-with-increased-blood-lipids-possible-cvd-risk/; Liu, B., Zhu, L., Wang, M., & Sun, Q. (2023). Associations between Per- and Polyfluoroalkyl Substances Exposures and Blood Lipid Levels among Adults—A Meta-Analysis. *Environmental Health Perspectives, 131*(5). https://doi.org/10.1289/ehp11840

[215]. Pitter, G., Jeddi, M. Z., Barbieri, G., Gion, M., Fabricio, A. S.,

Daprà, F., Russo, F., Fletcher, T., & Canova, C. (2020). Perfluoroalkyl substances are associated with elevated blood pressure and hypertension in highly exposed young adults. *Environmental Health, 19*(1). https://doi.org/10.1186/s12940-020-00656-0

[216]. Meneguzzi, A., Fava, C., Castelli, M., & Minuz, P. (2021). Exposure to perfluoroalkyl chemicals and cardiovascular disease: experimental and epidemiological evidence. *Frontiers in Endocrinology, 12*. https://doi.org/10.3389/fendo.2021.706352

[217]

. Steenland, K., Kugathasan, S., & Barr, D. B. (2018). PFOA and ulcerative colitis. *Environmental Research,* 165, 317–321. https://doi.org/10.1016/j.envres.2018.05.007

[218]. Qi, W., Clark, J. M., Timme Laragy, A. R., & Park, Y. (2020). Per- and polyfluoroalkyl substances and obesity, type 2 diabetes and non-alcoholic fatty liver disease: a review of epidemiologic findings. *Toxicological & Environmental Chemistry, 102*(1–4), 1–36. https://doi.org/10.1080/02772248.2020.1763997

[219]. Wikipedia contributors. (2024a, January 6). *Perfluorooctanesulfonic acid.* Wikipedia. Retrieved March 8, 2024, from https://en.wikipedia.org/wiki/Perfluorooctanesulfonic_acid

[220]. Wikipedia contributors. (2024c, February 12). *Perfluorooctanoic acid*. Wikipedia. Retrieved March 8, 2024, from https://en.wikipedia.org/wiki/Perfluorooctanoic_acid

[221]. Lerner, S. (2019, April 29). New teflon toxin causes cancer in lab animals. *The Intercept*. Retrieved March 8, 2024, from https://theintercept.com/2016/03/03/new-teflon-toxin-causes-cancer-in-lab-animals/

[222]. Li, F., Duan, J., Tian, S., Ji, H., Zhu, Y., Wei, Z., & Zhao, D. (2020). Short-chain per- and polyfluoroalkyl substances in aquatic systems: Occurrence, impacts and treatment. *Chemical Engineering Journal, 380*, 122506. https://doi.org/10.1016/j.cej.2019.122506; *Study: Newer PFAS chemicals 'May Pose More Risks' than those they replaced.* (2019, August 22). Environmental Working Group. Retrieved March 8, 2024,

from https://www.ewg.org/news-insights/news-release/study-newer-pfas-chemicals-may-pose-more-risks-those-they-replaced

[223]. *Flame retardants.* (n.d.-b). National Institute of Environmental Health Sciences. Retrieved March 9, 2024, from https://www.niehs.nih.gov/health/topics/agents/flame_retardants

[224]. *Polybrominated diphenyl ethers (PBDEs) and polybrominated biphenyls (PBBs) Factsheet | National Biomonitoring Program | CDC.* (n.d.). Retrieved March 9, 2024, from https://www.cdc.gov/biomonitoring/PBDEs_FactSheet.html

[225]. Israel, B. (2024b, February 20). *California to unveil new flammability standard to avoid chemicals in furniture.* Scientific American. Retrieved March 9, 2024, from https://www.scientificamerican.com/article/california-to-unveil-new-flammability-standard-to-avoid-chemicals-in-furniture/

[226]. Clean Production Action. (n.d.-b). *TV Reality: Toxic flame retardants in TVs | Clean Production Action Resources.* Retrieved March 9, 2024, from https://www.cleanproduction.org/resources/entry/report-toxic-flame-retardants-in-tvs

[227]. *Are flame retardants safe? Growing evidence says 'No.'* (n.d.). Yale E360. Retrieved March 9, 2024, from https://e360.yale.edu/features/pbdes_are_flame_retardants_safe_growing_evidence_says_no

[228]. Gibson, E. A., Siegel, E. L., Eniola, F., Herbstman, J. B., & Factor Litvak, P. (2018). Effects of polybrominated diphenyl ethers on child cognitive, behavioral, and motor development. *International Journal of Environmental Research and Public Health, 15*(8), 1636. https://doi.org/10.3390/ijerph15081636

[229]

. Xue, J., Xiao, Q., Zhang, M., Li, D., & Wang, X. (2023). Toxic effects and mechanisms of polybrominated diphenyl ethers. *International Journal of Molecular Sciences, 24*(17), 13487. https://doi.org/10.3390/ijms241713487

[230]. *Flame retardants.* (n.d.). UNEP - UN Environment Programme. Retrieved March 9, 2024, from https://www.unep.org/topics/

chemicals-and-pollution-action/pollution-and-health/persistent-organic-pollutants-pops/flame

[231]. *Phosphate Ester Flame Retardants | ToxFAQsTM | ATSDR*. (n.d.). Retrieved March 20, 2024, from https://wwwn.cdc.gov/TSP/ToxFAQs/ToxFAQsDetails.aspx?faqid=1164&toxid=239

[232]. *High levels of dioxins found in children's toys and other products made of recycled plastics Found in Argentina, Brazil, Cambodia, Canada, the EU, India, Japan and Nigeria | IPEN*. (n.d.). Retrieved March 9, 2024, from https://ipen.org/news/high-levels-dioxins-found-children%E2%80%99s-toys-and-other-products-made-recycled-plastics-found

[233]. Bača, D., Monroy, R., Castillo, M., Elkhazraji, A., Farooq, A., & Ahmad, R. (2023). Dioxins and plastic waste: A scientometric analysis and systematic literature review of the detection methods. *Environmental Advances, 13*, 100439. https://doi.org/10.1016/j.envadv.2023.100439

[234]. 7 Harmful chemical types in plastics. (2020). In *Endocrine Society*. Retrieved March 9, 2024, from https://www.endocrine.org/-/media/endocrine/files/topics/2020-dec-7-harmful-chemicals-backgrounder.pdf

[235]. Goldsberry, C. (2023, October 17). *Toxic flame retardants are still being used in plastics, even though there are alternatives*. Retrieved March 9, 2024, from https://www.plasticstoday.com/materials/toxic-flame-retardants-are-still-being-used-in-plastics-even-though-there-are-alternatives

[236]. *US Disposable straws - Industry Market Research - Market size, market share, market leaders, demand forecast, sales, company profiles, market Research, Industry Trends and Companies - The Freedonia Group*. (n.d.). The Freedonia Group. Retrieved March 14, 2024, from https://www.freedoniagroup.com/industry-study/us-disposable-straws

[237]. Perkins, T. (2023a, May 9). Many soft contact lenses in US made up of PFAS, research suggests. *The Guardian*. Retrieved March 11, 2024, from https://www.theguardian.com/environment/2023/may/09/contact-lenses-pfas-forever-chemicals

[238]. Zeeshan, M., Yang, Y., Zhou, Y., Huang, W., Wang, Z., Zeng, X., Liu, R. Q., Yang, B., Hu, L., Zeng, X. W., Sun, X., Yu, Y., & Dong, G. (2020). Incidence of ocular conditions associated with perfluoroalkyl substances exposure: Isomers of C8 Health Project in China. *Environment International, 137*, 105555. https://doi.org/10.1016/j.envint.2020.105555

[239]. Connor, N. (2023, July 24). *Melamine | Formula, Properties & Application*. Material Properties. Retrieved March 2, 2024, from https://material-properties.org/melamine/

[240]

. Wikipedia contributors. (2024, March 2). *Melamine foam*. Wikipedia. Retrieved March 2, 2024, from https://en.wikipedia.org/wiki/Melamine_foam#As_a_cleaner

[241]. Wikipedia contributors. (2024a, February 27). *Melamine*. Wikipedia. Retrieved March 2, 2024, from https://en.wikipedia.org/wiki/Melamine#Toxicity

[242]. WebMD Editorial Contributors. (2021, April 9). *What is melamine?* WebMD. Retrieved March 2, 2024, from https://www.webmd.com/a-to-z-guides/what-is-melamine

[243]. Nutrition, C. F. F. S. a. A. (2017, December 12). *Melamine in tableware questions and answers*. U.S. Food And Drug Administration. Retrieved March 2, 2024, from https://www.fda.gov/food/economically-motivated-adulteration-food-fraud/melamine-tableware-questions-and-answers

[244]. Center for Veterinary Medicine. (2018, September 4). *Melamine pet food recall of 2007*. U.S. Food And Drug Administration. Retrieved March 2, 2024, from https://www.fda.gov/animal-veterinary/recalls-withdrawals/melamine-pet-food-recall-2007

[245]. Gossner, C. M., Schlundt, J., Embarek, P. B., Hird, S., Lo-Fo-Wong, D., Beltran, J. J. O., Teoh, K. N., & Tritscher, A. (2009). The Melamine Incident: Implications for international food and feed safety. *Environmental Health Perspectives, 117*(12), 1803–1808. https://doi.org/10.1289/ehp.0900949

[246]. Staff Writer, with CNA. (2020, February 19). Officials caution

over melamine tableware items. *Taipei Times*. Retrieved March 2, 2024, from https://www.taipeitimes.com/News/taiwan/archives/2009/06/03/2003445238

[247]. Wu, C., Hsieh, T., Chen, B., Liu, C., & Wu, M. (2013). A crossover study of noodle soup consumption in melamine bowls and total melamine excretion in urine. *JAMA Internal Medicine, 173*(4), 317. https://doi.org/10.1001/jamainternmed.2013.1569; Hudson, K. (2023, September 7). The FDA says Melamine is Safe With One Big Caveat. *Organic Authority*. Retrieved March 2, 2024, from https://www.organicauthority.com/live-grow/why-avoid-melamine-dishes

[248]. Rachel. (2015, June 4). *How the Fashion Industry is Picking Up the Threads After Rana Plaza.* A\J. Retrieved March 16, 2024, from https://www.alternativesjournal.ca/community/how-the-fashion-industry-is-picking-up-the-threads-after-rana-plaza/

[249]. *Report 8/16: Hazardous chemical substances in textiles – proposals for risk management measures.* (n.d.). Kemikalieinspektionen. Retrieved March 16, 2024, from https://www.kemi.se/en/publications/reports/2016/report-8-16-hazardous-chemical-substances-in-textiles---proposals-for-risk-management-measures

[250]. Sandin, G., & Peters, G. (2018). Environmental impact of textile reuse and recycling – A review. *Journal of Cleaner Production, 184*, 353–365. https://doi.org/10.1016/j.jclepro.2018.02.266

[251]. Wikipedia contributors. (2024, January 29). *Polyester.* Wikipedia. Retrieved March 16, 2024, from https://en.wikipedia.org/wiki/Polyester

[252]

. McVeigh, K. (2021, November 29). Nurdles: the worst toxic waste you've probably never heard of. *The Guardian*. Retrieved March 12, 2024, from https://www.theguardian.com/environment/2021/nov/29/nurdles-plastic-pellets-environmental-ocean-spills-toxic-waste-not-classified-hazardous

[253]. Bourzac, K. (2024, February 6). *Grappling with the biggest marine plastic spill in history*. Chemical & Engineering News. Retrieved March 12, 2024, from https://cen.acs.org/environment/pollution/marine-plastic-spill-xpress-pearl-nurdle/101/i3

[254]. Van Der Molen, J., Van Leeuwen, S. M., Govers, L. L., Van Der Heide, T., & Olff, H. (2021). Potential Micro-Plastics dispersal and accumulation in the North Sea, with application to the MSC ZOE incident. *Frontiers in Marine Science, 8*. https://doi.org/10.3389/fmars.2021.607203

[255]. Sky News. (2018, February 24). South Africa's ecological 'nightmare' after plastic pellets spill. *Sky News*. Retrieved March 12, 2024, from https://news.sky.com/story/south-africas-ecological-nightmare-after-plastic-pellets-spill-11264554

[256]. Wikipedia contributors. (2023b, November 24). *Hong Kong plastic disaster*. Wikipedia. Retrieved March 12, 2024, from https://en.wikipedia.org/wiki/Hong_Kong_plastic_disaster

[257]. Wikipedia contributors. (2024, March 11). *East Palestine, Ohio, train derailment*. Wikipedia. https://en.wikipedia.org/wiki/East_Palestine,_Ohio,_train_derailment

[258]. *Ohio derailment: What chemicals spilled, and how could they affect residents?* (2023, February 22). NBC News. Retrieved March 12, 2024, from https://www.nbcnews.com/science/science-news/ohio-derailment-chemicals-spilled-impact-residents-rcna71561

[259]. Perkins, T. (2023, October 13). Toxic PFAS from US military bases polluting drinking water, report finds. *The Guardian*. Retrieved March 13, 2024, from https://www.theguardian.com/environment/2023/oct/13/pfas-pollution-us-military-bases-forever-chemicals

[260]. Perkins, T. (2023a, May 6). US agency takes unprecedented action to tackle PFAS water pollution. *The Guardian.* Retrieved March 13, 2024, from https://www.theguardian.com/us-news/2023/may/06/us-epa-pfas-drinking-water-pollution-ohio-river; *EPA takes first-ever federal Clean Water Act enforcement action to address PFAS discharges at Washington Works facility near Parkersburg, W. Va.*

| *US EPA*. (2023, April 26). US EPA. Retrieved March 13, 2024, from https://www.epa.gov/newsreleases/epa-takes-first-ever-federal-clean-water-act-enforcement-action-address-pfas

[261]. Perkins, T. (2023, October 27). New 'forever chemicals' polluting water near North Carolina plant, study finds. *The Guardian*. Retrieved March 8, 2024, from https://www.theguardian.com/environment/2023/oct/27/chemours-forever-chemicals-pfas-north-carolina-water-pollution

[262]. Frazin, R. (2024, February 23). United Nations criticizes 'forever chemical' contamination in North Carolina. *The Hill*. Retrieved March 13, 2024, from https://thehill.com/policy/energy-environment/4484124-united-nations-criticizes-forever-chemical-contamination-in-north-carolina/; Perkins, T. (2023c, November 28). UN human rights experts express alarm over PFAS pollution in North Carolina. *The Guardian*. Retrieved March 13, 2024, from https://www.theguardian.com/environment/2023/nov/28/north-carolina-pfas-pollution-united-nations-chemours-plant

[263]. *US companies DuPont and Chemours generated extensive contamination with toxic "forever chemicals" in North Carolina: UN experts*. (2024, February 21). [Press release]. Unite Nations Humans Rights Office of the High Commissioner. Retrieved March 8, 2024, from https://www.ohchr.org/en/press-releases/2024/02/us-companies-dupont-and-chemours-generated-extensive-contamination-toxic

[264]. Perkins, T. (2023b, November 17). US industry disposed of at least 60m pounds of PFAS waste in last five years. *The Guardian*. Retrieved March 13, 2024, from https://www.theguardian.com/environment/2023/nov/17/epa-pfas-forever-chemicals-waste-pollution-unregulated

[265]. Sargent, S. (2023, November 14). REPORT | *"Forever chemicals" disposal is creating a health nightmare (PDF)*. PEER.org. Retrieved March 13, 2024, from https://peer.org/report-forever-chemical-disposal/

[266]. Staff, E. (2023, July 5). *Fact Sheet: Single use plastics*. Earth Day. Retrieved March 15, 2024, from https://www.earthday.org/fact-

sheet-single-use-plastics/

[267]. Wikipedia contributors. (2024b, January 29). *Non-stick surface*. Wikipedia. Retrieved February 21, 2024, from https://en.wikipedia.org/wiki/Non-stick_surface

[268]. Elbein, S. (2024, February 28). Boiling tap water may be solution to microplastics. *The Hill*. Retrieved March 16, 2024, from https://thehill.com/policy/equilibrium-sustainability/4494416-boiling-tap-water-may-be-solution-to-microplastics/; Yu, Z., Wang, J., Liu, L., Liu, Z., & Zeng, E. Y. (2024). Drinking boiled tap water reduces human intake of nanoplastics and microplastics. *Environmental Science and Technology Letters*. https://doi.org/10.1021/acs.estlett.4c00081

[269]. Bosch, F. (2022, November 23). *9 in 10 Cosmetics Contain Microplastics*. Plastic Soup Foundation. Retrieved March 16, 2024, from https://www.plasticsoupfoundation.org/en/2022/04/almost-9-in-10-products-from-major-cosmetics-brands-contain-microplastics/

[270]. Safe Cosmetics. (2022, April 19). *Acrylates - safe cosmetics*. Retrieved March 16, 2024, from https://www.safecosmetics.org/chemicals/acrylates/

[271]. Burnett, C. L., Heldreth, B., Bergfeld, W. F., Belsito, D., Hill, R. A., Klaassen, C. D., Liebler, D. C., Marks, J. G., Shank, R. C., Slaga, T. J., Snyder, P. W., & Andersen, F. A. (2014). Safety assessment of nylon as used in cosmetics. *International Journal of Toxicology, 33*(4_suppl), 47S-60S. https://doi.org/10.1177/1091581814563524

[272]. Young, A. S., Allen, J. G., Kim, U., Seller, S. L., Webster, T. F., & Ceballos, D. (2018). Phthalate and organophosphate plasticizers in nail polish: Evaluation of labels and ingredients. *Environmental Science & Technology, 52*(21), 12841–12850. https://doi.org/10.1021/acs.est.8b04495

[273]. Nutrition, C. F. F. S. a. A. (2024, January 22). *Per and Polyfluoroalkyl Substances (PFAS) in Cosmetics*. U.S. Food And Drug Administration. Retrieved March 16, 2024, from https://www.fda.gov/cosmetics/cosmetic-ingredients/and-polyfluoroalkyl-substances-pfas-cosmetics

[274]. Learn, J. R. (2023, November 22). Saying goodbye to glitter? What to know about Europe's glitter ban. *Discover Magazine*. Retrieved March 16, 2024, from https://www.discovermagazine.com/environment/saying-goodbye-to-glitter-what-to-know-about-europes-glitter-ban
[275]. Goff, K. (2022, August 22). How bad are the chemicals in your mattress? *CNET*. Retrieved March 16, 2024, from https://www.cnet.com/health/sleep/how-bad-are-the-chemicals-in-your-mattress/
[276]. Rankin, S. (2021, July 16). *Mushroom packaging - the plastic of the future*. Natura Mushrooms. Retrieved March 16, 2024, from https://naturamushrooms.com/blogs/news/mushroom-packaging-the-plastic-of-the-future; *Our process*. (n.d.). Mushroom Packaging. Retrieved March 16, 2024, from https://mushroompackaging.com/pages/our-process
[277]. *Nations sign up to end global scourge of plastic pollution*. (2022, March 4). UN News. Retrieved March 16, 2024, from https://news.un.org/en/story/2022/03/1113142
[278]. Charlton, A., & McDERMOTT, J. (2023, June 3). Delegates working to end global plastics pollution agree to craft a draft treaty. *AP News*. Retrieved March 16, 2024, from https://apnews.com/article/plastic-pollution-treaty-un-paris-talks-c9728d632bdcf505f40224bedf07a8e1
[279]. *Turning off the Tap: How the world can end plastic pollution and create a circular economy*. (n.d.). UNEP - UN Environment Programme. Retrieved March 16, 2024, from https://www.unep.org/resources/turning-off-tap-end-plastic-pollution-create-circular-economy; Carrington, D. (2023, May 16). Plastic pollution could be slashed by 80% by 2040, UN says. *The Guardian*. Retrieved March 16, 2024, from https://www.theguardian.com/environment/2023/may/16/plastic-pollution-could-be-slashed-by-80-by-2040-un-says
[280]. Lean, G. (2023, June 13). Whisper it, but the boom in plastic production could be about to come to a juddering halt. *The Guardian*. Retrieved March 16, 2024, from https://www.theguardian.com/

commentisfree/2023/jun/13/plastic-production-plastics-treaty-environmental-diplomacy